VOLUME ONE HUNDRED AND FOURTEEN

ADVANCES IN VIRUS RESEARCH

Viruses and Climate Change

VOLUME ONE HUNDRED AND FOURTEEN

ADVANCES IN VIRUS RESEARCH

Viruses and Climate Change

Edited by

MARILYN J. ROOSSINCK

Department of Plant Pathology and Environmental Microbiology, Center for Infectious Disease Dynamics, Penn State University, University Park, PA, United States

ELSEVIER

ACADEMIC PRESS

An imprint of Elsevier

Academic Press is an imprint of Elsevier
50 Hampshire Street, 5th Floor, Cambridge, MA 02139, United States
525 B Street, Suite 1650, San Diego, CA 92101, United States
The Boulevard, Langford Lane, Kidlington, Oxford OX5 1GB, United Kingdom
125 London Wall, London, EC2Y 5AS, United Kingdom

First edition 2022

ISBN: 978-0-323-91212-9
ISSN: 0065-3527

For information on all Academic Press publications
visit our website at https://www.elsevier.com/books-and-journals

Publisher: Zoe Kruze
Acquisitions Editor: Leticia Lima
Developmental Editor: Federico Paulo S. Mendoza
Production Project Manager: James Selvam
Cover Designer: Vicky Pearson

Typeset by STRAIVE, India

Contents

Contributors *vii*
Preface *ix*

1. Challenges and opportunities for plant viruses under a climate change scenario **1**
Nuria Montes and Israel Pagán

1. Introduction 2
2. Climate change and plant virus pathogenicity 8
3. Climate change and plant virus transmission 16
4. Climate change and plant virus ecology 27
5. Climate change and plant virus evolution 33
6. Climate change and management of plant virus diseases 38
7. Concluding remarks and future perspectives 44

Acknowledgments 48
References 48

2. Marine viruses and climate change: Virioplankton, the carbon cycle, and our future ocean **67**
Hannah Locke, Kay D. Bidle, Kimberlee Thamatrakoln, Christopher T. Johns, Juan A. Bonachela, Barbra D. Ferrell, and K. Eric Wommack

1. Introduction 68
2. Climate change effects on the global ocean 74
3. Key virus–host players in the marine carbon cycle 81
4. Modern approaches to investigating virus–host dynamics in a changing climate 109
5. Overall takeaways and conclusions 121

Acknowledgments 122
References 122

3. West Nile virus and climate change **147**
Rachel L. Fay, Alexander C. Keyel, and Alexander T. Ciota

1. Introduction 148
2. Temperature, viral fitness and vector competence for West Nile virus 153
3. Mosquito physiology and climate 160
4. Epidemiological models of West Nile virus and climate 164

5. The influence of temperature on West Nile virus diversity and evolution 172
6. Concluding remarks 174
References 176
Further reading 193

Contributors

Kay D. Bidle
Rutgers Univ., Dept. of Marine & Coastal Sciences, New Brunswick, NJ, United States

Juan A. Bonachela
Rutgers Univ., Dept. of Ecology, Evolution & Natural Resources, New Brunswick, NJ, United States

Alexander T. Ciota
The Arbovirus Laboratory, Wadsworth Center, New York State Department of Health, Slingerlands; Department of Biomedical Sciences, State University of New York at Albany School of Public Health, Rensselaer, NY, United States

Rachel L. Fay
The Arbovirus Laboratory, Wadsworth Center, New York State Department of Health, Slingerlands; Department of Biomedical Sciences, State University of New York at Albany School of Public Health, Rensselaer, NY, United States

Barbra D. Ferrell
Univ. of Delaware, Delaware Biotechnology Inst., Newark, DE, United States

Christopher T. Johns
Rutgers Univ., Dept. of Marine & Coastal Sciences, New Brunswick, NJ, United States

Alexander C. Keyel
The Arbovirus Laboratory, Wadsworth Center, New York State Department of Health, Slingerlands; Department of Atmospheric and Environmental Sciences, State University of New York at Albany, Albany, NY, United States

Hannah Locke
Univ. of Delaware, Delaware Biotechnology Inst., Newark, DE, United States

Nuria Montes
Fisiología Vegetal, Departamento Ciencias Farmacéuticas y de la Salud, Facultad de Farmacia, Universidad San Pablo-CEU Universities; Servicio de Reumatología, Hospital Universitario de la Princesa, Instituto de Investigación Sanitaria (IIS-IP), Madrid, Spain

Israel Pagán
Centro de Biotecnología y Genómica de Plantas UPM-INIA and E.T.S. Ingeniería Agronómica, Alimentaria y de Biosistemas, Universidad Politécnica de Madrid, Madrid, Spain

Kimberlee Thamatrakoln
Rutgers Univ., Dept. of Marine & Coastal Sciences, New Brunswick, NJ, United States

K. Eric Wommack
Univ. of Delaware, Delaware Biotechnology Inst., Newark, DE, United States

Preface

The effects of climate change loom over almost every aspect of our lives, from changes in weather patterns to changes in agriculture, access to food and water, and changing demographics in worldwide populations. In this volume, we look at the effects of climate change on viruses. Chapter 1 explores what climate change may mean for plant viruses. Many plant viruses that infect crops are temperature-sensitive and do not replicate well under high temperatures, so climate change may ameliorate virus disease in some crops. On the other hand, climate change is leading to changes in insect distribution, and as major vectors for plant viruses the spread of viruses to new regions due to increased insect loads may be significant.

In Chapter 2, we take a deep dive into the impacts of climate change on marine viruses. The complex and critical role of viruses in maintaining the ecology of the oceans will certainly be impacted by warming oceans, changing oceans currents, and changes in the microbe populations that are the most common hosts for marine viruses. These changes could impact the ecology of the entire planet, as viruses are critical for maintaining carbon balance and nutrient cycles.

In Chapter 3, we look at the impact of climate change on an insect-transmitted virus that infects humans as a dead-end host. The range of many insect vectors is changing with the warming planet, leaving new areas vulnerable to insect-borne diseases that were previously restricted by ambient temperatures.

These chapters pose more questions than they can answer at this time. Adequate research is lacking in many places, and the results of climate change remain to be seen. Models, however, can predict what is likely to happen in some cases, and the world needs to prepare for the impacts of climate change on virus-related diseases and global ecological cycles.

CHAPTER ONE

Challenges and opportunities for plant viruses under a climate change scenario

Nuria Montes[a,b] and Israel Pagán[c,*]

[a]Fisiología Vegetal, Departamento Ciencias Farmacéuticas y de la Salud, Facultad de Farmacia, Universidad San Pablo-CEU Universities, Madrid, Spain

[b]Servicio de Reumatología, Hospital Universitario de la Princesa, Instituto de Investigación Sanitaria (IIS-IP), Madrid, Spain

[c]Centro de Biotecnología y Genómica de Plantas UPM-INIA and E.T.S. Ingeniería Agronómica, Alimentaria y de Biosistemas, Universidad Politécnica de Madrid, Madrid, Spain

*Corresponding author: e-mail address: jesusisrael.pagan@upm.es

Contents

1. Introduction 2
2. Climate change and plant virus pathogenicity 8
3. Climate change and plant virus transmission 16
  3.1 Horizontal transmission 16
  3.2 Vertical transmission 24
4. Climate change and plant virus ecology 27
  4.1 Host and geographic range 28
  4.2 Mixed infections 31
5. Climate change and plant virus evolution 33
6. Climate change and management of plant virus diseases 38
  6.1 Detection and diagnosis 38
  6.2 Chemical and biological control of vectors 40
  6.3 Genetic resistance and tolerance 41
  6.4 Other methods 43
7. Concluding remarks and future perspectives 44
Acknowledgments 48
References 48

Abstract

There is an increasing societal awareness on the enormous threat that climate change may pose for human, animal and plant welfare. Although direct effects due to exposure to heat, drought or elevated greenhouse gasses seem to be progressively more obvious, indirect effects remain debatable. A relevant aspect to be clarified relates to the relationship between altered environmental conditions and pathogen-induced diseases. In the particular case of plant viruses, it is still unclear whether climate change will primarily represent an opportunity for the emergence of new infections in previously

Advances in Virus Research, Volume 114
ISSN 0065-3527
https://doi.org/10.1016/bs.aivir.2022.08.001

uncolonized areas and hosts, or if it will mostly be a strong constrain reducing the impact of plant virus diseases and challenging the pathogen's adaptive capacity. This review focuses on current knowledge on the relationship between climate change and the outcome plant-virus interactions. We summarize work done on how this relationship modulates plant virus pathogenicity, between-host transmission (which include the triple interaction plant-virus-vector), ecology, evolution and management of the epidemics they cause. Considering these studies, we propose avenues for future research on this subject.

1. Introduction

A central paradigm of Plant Pathology states that the existence of a disease caused by a biotic agent requires the interaction of a susceptible host, a virulent pathogen, and an environment favorable for disease development in what is called the Disease Triangle (Agrios, 2005), which becomes a pyramid for pathogens transmitted through vectors (Fig. 1). Although plant pathologists have traditionally devoted more attention to the first two vertices of the triangle, accumulating evidence indicates that the environment plays a central role in plant disease dynamics (Amari et al., 2021; Garrett et al., 2006; Jones, 2016; Pagán et al., 2012). Environmental cues include climatic conditions, and much of the interest in the third vertex of the Disease Triangle derives from the increasing awareness of the current context of climate change. Consequently, in the last 20 years the number of research articles on this topic has exponentially increased (Fig. 2). Climate change is a multifaceted phenomenon that entails, among other things, increasing concentrations of atmospheric greenhouse gases (particularly CO_2), rising temperatures, changes in precipitation patterns, and higher light intensity (IPCC, 2014). In the last decades, the rate at which climate change occurs has accelerated. Indeed, atmospheric CO_2 increased from 310 to 420 ppm in the last 65 years, and it is predicted to reach 730–1000 ppm by the year 2100 (Collins et al., 2013; IPCC, 2014). In the same period, global average surface temperature raised 0.6 °C. If this trend continues until the end of the century, global warming will reach an additional 1.0–3.7 °C (Collins et al., 2013; IPCC, 2014). At these elevated levels of CO_2 and temperature, stratocumulus decks would break, leading to reduced cloud coverage and increased light intensity (Schneider et al., 2019). These climatic alterations are thought to result in an increased frequency, intensity, and duration of extreme weather events in most land regions, including intensity of heavy precipitation and frequency of flooding, frequency and intensity of droughts,

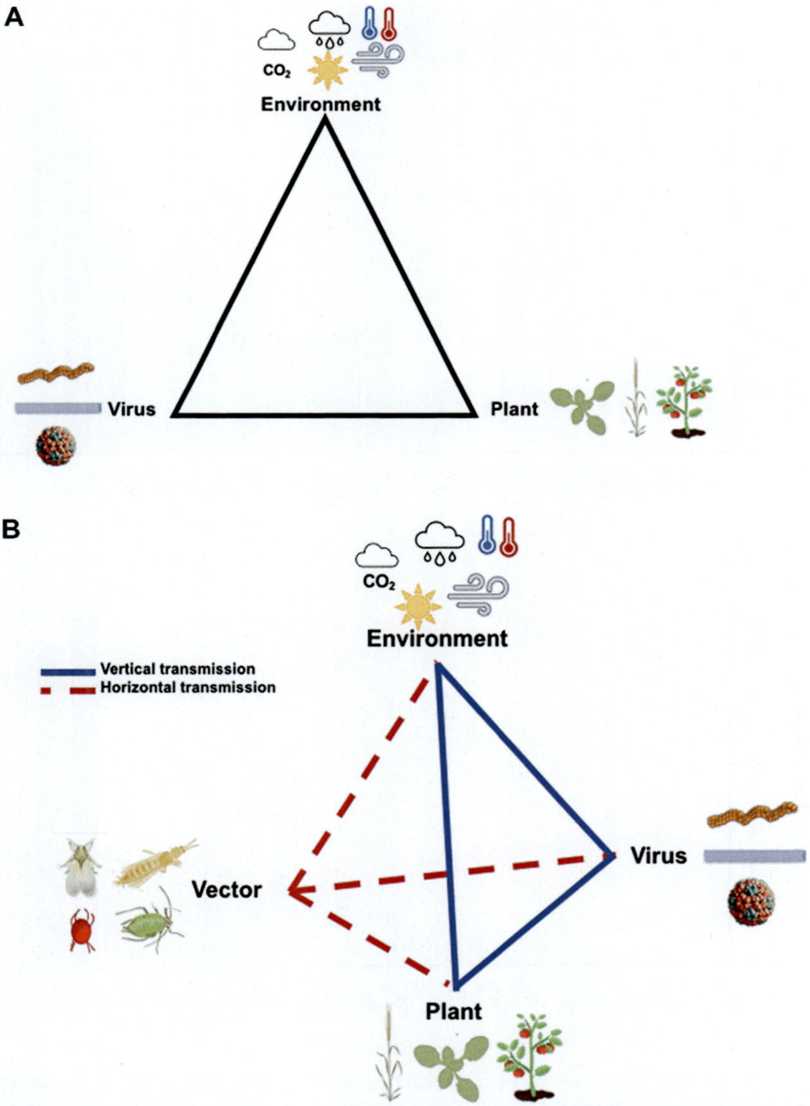

Fig. 1 Disease triangles for plant-virus interactions without (A) and with (B) the involvement of virus vectors. Lines indicate two-way interactions between virus, host, and environment (A) and virus, host, environment and vector (B). In the context of this review, environment can be approximated to climate change and the factors typically associated with this phenomenon (altered temperature, CO_2, light intensity, rainfall and wind speed) are represented. In (B), dashed red lines indicate two-way relationships affecting horizontal transmission, and solid blue lines indicate two-way relationships affecting vertical transmission.

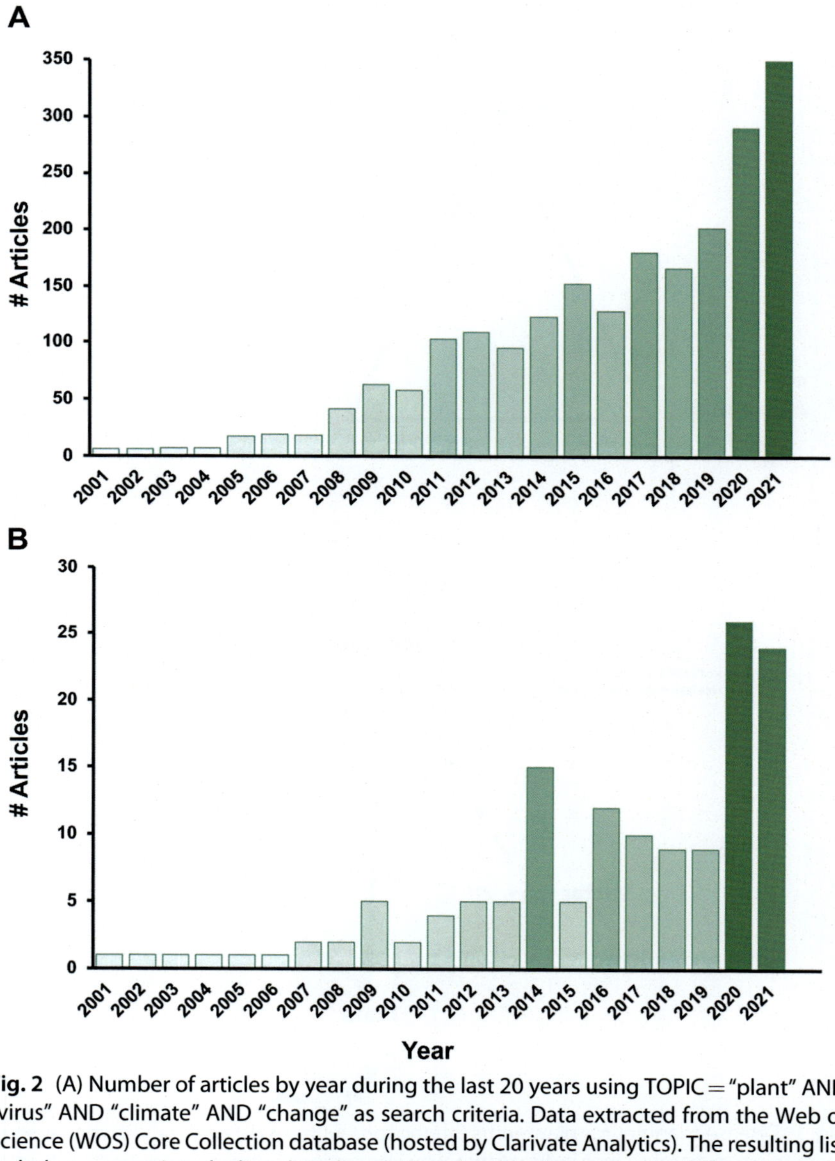

Fig. 2 (A) Number of articles by year during the last 20 years using TOPIC = "plant" AND "virus" AND "climate" AND "change" as search criteria. Data extracted from the Web of Science (WOS) Core Collection database (hosted by Clarivate Analytics). The resulting list includes manuscripts dealing directly or indirectly with climate change (effect of a single abiotic stress associated with climate change on any aspect plant-virus interactions, including plant and vector biology). (B) Number of articles by year using the list obtained in (A) and filtering for those directly studying plant-virus interactions in the context of climate change.

desertification in some dryland areas and alterations of windspeed (dust storms, tornados, hurricanes, etc.) (Ebi et al., 2021). Since these climatic alterations may influence most (if not all) biological processes, climate change is expected to have a huge impact in the reproductive success of living organisms and in the relationships that they establish (e.g., Grimm et al., 2013; Jennings and Harris, 2017), and plant-virus interactions are not an exception (Jones, 2016; Trębicki, 2020; Velásquez et al., 2018).

Many reviews have dealt with the effects of climate change on plant pathogens, but plant viruses have been comparatively less represented (Jones, 2016; Trębicki, 2020). For example, an overview of 75 review papers on this topic published during the period 1988–2019 revealed that only 6 explicitly considered plant viruses (Jeger, 2020; Juroszek et al., 2020). Similarly, a review of 70 mathematical models projecting disease risk in a climate change scenario only included 2 focused on plant viruses (Juroszek and Tiedemann, 2015) (Fig. 2). Despite their paucity, these review papers generally agree on the key role that climate change may play for many plant virus traits (Canto et al., 2009; Jones, 2016; Trębicki, 2020). However, they also show frequent contradictory results on the sign and magnitude of such effects. Hence, one of the main questions that remains to be answered is whether the expected effects of climate change will be favorable or detrimental for plant viruses, and whether pleiotropic effects can be expected according to the context in which climatic alterations occur (climatic factors, plant-virus-vector interaction and ecosystem involvement), which may explain the observed contradictions.

Negative effects of climate change that can be categorized as direct (those that modify the virus life cycle) or indirect (affecting the biology of their hosts/vectors) (Jones, 2016), may pose an enormous challenge for plant viruses. These effects can be summarized as:

- Climate change may directly alter virus-controlled traits. For instance, higher temperatures may reduce virion stability (Chakravarty et al., 2020), which is crucial for survival outside the host, particularly for viruses transmitted by contact (Orlov et al., 1998). Also, altered climate may reduce virus within-host multiplication and virulence (Canto and Palukaitis, 2002; Zhang et al., 2012), which may have negative consequences for virus fitness (Anderson and May, 1982).
- Climate change may alter the host physiology in such way that it reduces virus fitness. For instance, induction of stronger plant defenses may prevent infection, potentially leading to virus extinction. Certain climatic conditions may reduce plant attractiveness to vectors or competence (i.e., the efficiency of the vector to transmit the virus) (Jones and Barbetti, 2012).

- Altered climate threatens plant welfare. It has been estimated that, at its current rate, climate change would lead to the extinction of one in six species in the next few decades (Urban, 2015), mostly due to increased exposure to heat (Román-Palacios and Wiens, 2020; Vicedo-Cabrera et al., 2021). Because plant viruses intimately depend on their hosts for survival, such changes on ecosystem composition may indirectly, but severely, affect plant viruses: local or global disappearance of susceptible plant populations due to altered climate may strongly restrict the chances for virus transmission, and hence for survival. This may me particularly important for highly specialized viruses with narrow host ranges (Jones, 2016), for which the lack of the susceptible host may drive the population to extinction. A similar rationale can be applied to the effect of climate change on virus vectors (Canto et al., 2009).
- Even if climate change does not lead to the extinction of hosts and vectors, it may cause strong changes in their biology. Altered phenology of plants and vectors may result in desynchronization of their population dynamics such that peaks of virus infection do not coincide with vector presence, leading to reduced or no transmission. For example, elevated temperatures may result in earlier growth of vector populations, such that at highest vector density the host plants are still young or have not germinated at all. Conversely, early flowering may lead to an acceleration of plant senescence, such that plants die before vector maximum density, decoupling transmission (Culbreath and Srinivasan, 2011).

On the other hand, climate change has potential benefits for plant viruses opening new opportunities for disease emergence. Increasing effort has been devoted to understanding these effects as they are the more concerning for plant and ecosystem health. For instance, several mathematical models highlighted that climate modifications associated with global warming result in increasing severity, distribution and spread of virus diseases (reviewed by Jeger et al., 2018). Briefly, expected positive consequences of climate change are:

- Climate change may affect pathogen virulence and/or host susceptibility. For instance, higher temperatures may render plant effector-triggered immunity ineffective (Wang et al., 2009; Whitham et al., 1996). Combined heat and drought can enhance the negative effects of pathogen infections on plant growth (Prasch and Sonnewald, 2013) and elevated CO_2 may increase pathogen load (Del Toro et al., 2015, 2017; Trębicki et al., 2016).

- Climate change may create favorable conditions for plant pathogen growth in locations which were previously unfavorable, enlarging the pathogen's distribution area and leading to disease emergence (Bosso et al., 2016; Fones and Gurr, 2017; Jones, 2016). Similarly, it has also far-reaching consequences for pathogens transmitted through vectors by inducing alterations in the vector population dynamics and behavior that enhance horizontal transmission (Canto et al., 2009; Jamieson et al., 2017; Wu et al., 2020). Indeed, drought and higher temperature positively affect plant virus transmission by increasing vector reproductive success and flying activity, and virus acquisition/release (Culbreath and Srinivasan, 2011; Diaz and Fereres, 2005; van Munster et al., 2017). Also, elevated temperature and CO_2 increase aphid feeding rates (Sun et al., 2016), which favors virus transmission (Fereres and Moreno, 2009; Wamonje et al., 2020). Climate change may also increase the chances for transmission, for instance by increasing virus multiplication rate (Chung et al., 2015; van Munster et al., 2017) or plant attractiveness to vectors (Jones and Barbetti, 2012).
- Mirroring negative effects, climate change may expand the lifespan of host plants and vectors such that populations previously isolated due to seasonal effects are brought to overlap. This synchrony may eventually render greater availability of new vectors, increasing the chances for virus transmission (Anwar et al., 2022). Similarly, climate change may affect the seasonal overlapping of vectors and their natural enemies, also affecting the chances for virus transmission (Macfadyen et al., 2018; Moreno-Delafuente et al., 2021; Skendžic et al., 2021).
- Finally, altered climatic conditions may promote vector dispersal. For instance, stronger winds may enlarge the vector dispersion area, and altered temperatures may allow vectors to establish in locations previously unfavorable for them (Jones, 2016; Kirchner et al., 2013).

In sum, climate change can be a two-edged sword for plant viruses. On one hand, it may challenge their adaptive capacity, putting at risk virus survival. On the other hand, it provides opportunities to expand their distribution, and increase their transmission rate and prevalence. In this review, we aim at summarizing relevant research supporting both positive and negative impacts of climate change on plant viruses. We have structured the chapter around five major aspects for which the effect of climate change on plant virus populations has been studied: Pathogenicity, transmission, ecology, evolution and control.

2. Climate change and plant virus pathogenicity

From the human perspective, the most relevant effect of the environment on plant-virus interactions is how it modulates disease development due to infection. This is because pathogenic effects of virus presence may have a strong negative impact on the yield of cultivated plants. Consequently, a large body of experimental work has dealt with the interaction between climatic variables and plant virus infectivity and pathogenicity (generally measured as symptom severity) (Fig. 3).

To cause a disease, the first requirement for the pathogen is infecting the host. For plant viruses, the majority of the work on this aspect focused on the effect of environmental conditions on the protection conferred by major resistance genes (R genes). As many of these are temperature-sensitive, climate change may alter their effectiveness and durability. For instance, wheat plants harboring the *Wsm1* resistance gene inoculated with *Wheat streak mosaic virus* (WSMV) lost protection against virus infection at 20–25 °C (Wosula et al., 2017). However, in the absence of R genes, infectivity can also be controlled by temperature. In the same work, susceptible wheat plants maintained at 10 °C restricted WSMV to the entry points, whereas at

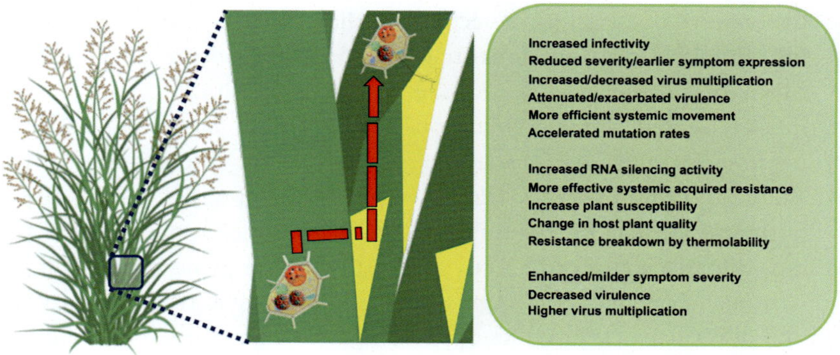

Fig. 3 Schematic representation (left) and summary (right) of reported effects of climate change on plant virus infectivity, pathogenicity and within-host multiplication/movement. Scheme shows in the left side the effects on virus infectivity (solid red arrow) linked to thermolability of resistance genes (yellow triangle); and in the right side effects on systemic movement (red dashed arrow), RNA silencing (blue and yellow double helix), multiplication (spheric virions) and symptoms (yellow leaf areas). Summary is divided in three blocks: top, effects on virus trait; middle, effects on plant traits; bottom, effects associated with mixed infections. Scheme was constructed using BioRender.

temperatures above 15 °C the virus moved systemically. Similarly, systemic infection of *Potato virus Y* (PVY) in susceptible potato plants was observed only within the temperature range of 16–32 °C, but not above or below this range (Choi et al., 2017). The effect of climate change on plant infectivity will be more extensively discussed in Section 6 in the context of control strategies against virus infections. Once the virus infects the plant systemically, theoretical elaborations propose that the effect of infection on plant growth and reproduction depend on the level of virus multiplication within the host (Anderson and May, 1982; Hull, 2014), although this is not necessarily so for all plant-virus interactions (Pagán et al., 2007, 2008). Environmental conditions associated with climate change (elevated temperature, CO_2 or light intensity, and drought) may alter both symptom severity and virus multiplication, as well as the relationship between these traits. Hence, we will discuss together the existing analyses considering the effect of climate change in virus multiplication and pathogenicity.

That temperature modulates disease development and virus titer has been known for more than a century (Johnson, 1921). In this experiment, tobacco plants infected by TMV developed mottle symptoms at 20 °C that were milder at 30 °C and disappeared either at 36 °C or at 10 °C, indicating an optimal range of temperatures for symptom development coinciding with the optimal physiological temperature for plant growth. Accordingly, various studies established that TMV optimal multiplication was around 30 °C, with much lower titers in conditions above 35 °C or below 13 °C (Harrison, 1956; Yarwood, 1952). In the late 1950s, a review on this subject cited several works, most of which reflected a trend toward reduced symptom severity and virus multiplication at higher temperatures (Kassanis, 1957). In this line, recent analyses points to the same trend. For instance, *Cymbidium ringspot virus* infection in *Nicotiana benthamiana* induced more severe symptoms at 15 °C, which was correlated with the inhibition of the RNA silencing response of the plant, than at 27 °C (Szittya et al., 2003). Also, *Plum pox virus* (PPV) accumulated to lower levels and induced milder symptoms at 30 °C than at 25 °C (Aguilar et al., 2015). Nancarrow et al. (2014) showed that elevated temperature resulted in milder effects of *Barley yellow dwarf virus* (BYDV) infection on wheat growth, which was associated with lower BYDV accumulation at late times postinfection. However, Kassanis (1957) has already noted that the effect of temperature depended on the virus isolate-plant genotype-temperature combination. Indeed, numerous cases of increased pathogenicity at extreme temperatures have been reported. For example, infection by a Japanese isolate of *Rehmannia mosaic virus* leads to mottle in tomato when incubated at 20 °C and induces systemic necrosis at 25 °C (Hamada et al., 2019). Also, *Turnip mosaic virus* (TuMV) infection induced more severe symptoms in Chinese cabbage at higher temperatures, which correlated with higher virus multiplication (Chung et al., 2015). Higher temperatures may not only result in more severe symptoms but also in faster development of symptoms, which again may be linked to accelerated kinetics of virus accumulation as shown in the *Peanut stunt virus-N. benthamiana* (Obrepalska-Steplowska et al., 2015) and *Capsicum chlorosis virus*-pepper (Tsai et al., 2022) interactions.

The effect of elevated CO_2 (eCO_2), the other major consequence of climate change (Canto et al., 2009), on plant virus pathogenicity also has received considerable attention. Elevated CO_2 resulted in attenuated symptoms of *Potato virus X* (PVX) in *N. benthamiana*, which correlated with a lower virus accumulation (Aguilar et al., 2015). Also, eCO_2 decreased *Cucumber mosaic virus* (CMV) copy number in tobacco and increased calcium concentration and expression of the calcium-binding protein rgs-CaM in infected plants, which directly weakened the function of the 2b protein, the viral silencing suppressor. These changes correlated with lesser effects of infection on plant biomass (Guo et al., 2021). Similarly, eCO_2 resulted in milder symptoms of PVY infection in tobacco and of *Tomato yellow leaf curl virus* (TYLCV) in susceptible tomatoes (Guo et al., 2016; Ye et al., 2010). Although virus multiplication was not measured in these studies, milder virus infection in tomato was accompanied by stronger activation of the SA-signaling pathway, which would induce a reduction of virus titer. This effect was genotype specific and in tomatoes harboring the *Mi-1.2* gene that confers resistance against nematodes, eCO_2 enhanced TYLCV symptoms and repressed the SA-signaling pathway (Guo et al., 2016). Notably, examples of intermediate effects between these two extremes also exist. For example, Trębicki et al. (2015) showed that eCO_2 significantly increased BYDV multiplication in wheat, but this had no effect on plant growth. These authors explained this result by the positive effect of eCO_2 in plant growth and development that would confer tolerance to infection. Overall, although a reduction of virus multiplication and symptom severity seem to be the most prevalent effects, these studies highlight not only the variety of outcomes in the interaction between eCO_2 and plant virus infections, but also the diversity of molecular mechanisms that may determine the results of such interactions.

In line with the major effects observed for eCO_2, high light intensity increased *N. benthamiana* RNA silencing, reducing *Tobacco rattle virus* symptoms and multiplication level (Kotakis et al., 2011). Similarly, preinoculation exposure of tomato plants to UV light effectively suppressed *Tomato mosaic virus* infection symptoms, with postinoculation exposure resulting in much milder symptoms than in nonexposed ones (Matsuura and Ishikura, 2014). In contrast, *Arabidopsis thaliana* plants preexposed to UV light were much more susceptible to *Oilseed rape mosaic virus* infection than nonexposed plants (Yao et al., 2012). This virus-specific (and even isolate-specific) interaction with light was also observed by Montes and Pagán (2019), who reported that *A. thaliana* plants subjected to high light intensity were more susceptible to UK 1-TuMV, but more resistant (and suffer less from infection) to JPN 1-TuMV and LS-CMV, than plants subjected to low light intensity. Interestingly, Yao et al. (2012) showed that *A. thaliana* plants placed in the vicinity of UV-exposed plants became more resistant to virus infection, suggesting some type of "information exchange" among plants, perhaps through volatiles. These results add another layer of complexity to the effects of climate change in plant-virus interactions pointing to phenotypic plasticity of plants nonexposed to an abiotic stress through plant-to-plant communication. Higher light intensity also affected the effectiveness of other plant defenses. For instance, plant tolerance (see Section 6) to UK 1-TuMV infection was less effective at higher than at lower light intensity, whereas the opposite was observed for JPN 1-TuMV and LS-CMV (Montes and Pagán, 2019).

Another major expected effect of climate change is the alterations in water availability. On the one hand, higher average temperature is expected to increase the frequency of dry years and therefore the frequency, magnitude, and duration of drought events (Diffenbaugh et al., 2015). *A. thaliana* plants simultaneously subjected to drought and TuMV infection showed higher mortality rates than mock-inoculated drought stressed controls, due to virus-induced alterations of the drought-responsive gene RD29A (Manacorda et al., 2021). In 24 natural *A. thaliana* genotypes, *Cauliflower mosaic virus* (CaMV) virulence generally increased under water deficit while viral within-host accumulation was significantly altered in only a few plant genotypes (Bergès et al., 2021). However, increased virus virulence under drought is not a general rule. González et al. (2021) showed that, in *A. thaliana*, TuMV infection exhibited increased or reduced virulence under drought in a host genotype-specific manner: in plant genotypes showing lower TuMV virulence at standard watering conditions, virus evolution led to a larger increase of the disease progression curves under drought than under standard conditions, whereas the opposite was observed in host genotypes more susceptible to TuMV in the absence of water deprivation. In line with this later group of *A. thaliana* genotypes, *Grapevine fanleaf virus* (GFLV)-infected vines showed milder symptoms under mild (but not severe) water stress (Jež-Krebelj et al., 2022). On the other hand, increased frequency of extreme weather events due to climate change may result in more frequent flooding (Brunner et al., 2021). However, to our knowledge specific work on the interaction between flooding and plant virus infection has not been done to date (Martínez-Arias et al., 2022).

The work above addressed studies in which a single factor linked to climate change is altered, although in nature plant viruses will be subjected to various abiotic stresses simultaneously. Hence, understanding the interaction between different conditions linked to climate change in determining the outcome of plant virus infections is of great importance. Although not frequent, some analyses addressed such interactions. For instance, the combined effect of higher temperature and eCO_2 was analyzed on *N. benthamiana* plants infected by PVX, PVY and CMV. These conditions reduced PVX and PVY, but not CMV, multiplication and pathogenicity (Del Toro et al., 2017; Makki et al., 2021). These results seemed to be due to changes in plant chemistry but not to alterations of the RNA silencing plant defense (Del Toro et al., 2015). Similar to CMV, *Potato leafroll virus* (PLRV) infection in potato was unaffected by combined heat and eCO_2 (Chung et al., 2017), whereas similar to PVX and PVY, reductions of PPV multiplication and pathogenicity were observed in *N. benthamiana* and *A. thaliana* subjected to the same combination of abiotic stresses (Aguilar et al., 2020). In this latter work, the effect of drought in combination with higher temperature and eCO_2 on virus pathogenicity also was analyzed, showing more deleterious effects than in plants grown at standard conditions. Combined heat and drought also resulted in higher TuMV multiplication and pathogenicity in *A. thaliana* (Prasch and Sonnewald, 2013). These works highlight that the study of individual effects of each abiotic stress do not allow prediction of their combined effects (Table 1), which underscores the need of more work addressing the simultaneous effect of conditions linked to climate change on plant-virus interactions.

Table 1 Summary of studies on the effect of abiotic stresses associated with climate change on plant virus within-host multiplication and pathogenicity.

Abiotic stress(es)	Virus	Host	Multiplication	Pathogenicity	Reference
Temperature	*Barley yellow dwarf virus* (BYDV)	*Triticum aestivum*	–	–	Nancarrow et al. (2014)
	Capsicum chlorosis virus	*Capsicum annuum*	+	+	Tsai et al. (2022)
	Cymbidium ringspot virus	*Nicotiana benthamiana*	ND	–	Szittya et al. (2003)
	Peanut stunt virus	*N. benthamiana*	+	+	Obrepalska-Steplowska et al. (2015)
	Plum pox virus (PPV)	*N. benthamiana*	–	–	Aguilar et al. (2015)
	Rehmannia mosaic virus	*Solanum lycopersicum*	ND	+	Hamada et al. (2019)
	Tobacco mosaic virus	*Nicotiana tabacum*	–	–	Johnson (1921); Harrison (1956); Yarwood (1952)
	Turnip mosaic virus (TuMV)	*Brassica campestris*	+	+	Chung et al. (2015)
CO_2	BYDV	*T. aestivum*	+	o	Trębicki et al. (2015)
	Cucumber mosaic virus (CMV)	*N. tabacum*	–	–	Guo et al. (2021)

Continued

Table 1 Summary of studies on the effect of abiotic stresses associated with climate change on plant virus within-host multiplication and pathogenicity.—cont'd

Abiotic stress(es)	Virus	Host	Multiplication	Pathogenicity	Reference
	Potato virus X (PVX)	*N. benthamiana*	–	–	Aguilar et al. (2015)
	Potato virus Y (PVY)	*N. tabacum*	ND	–	Ye et al. (2010)
	Tomato yellow leaf curl virus (TYLCV)	*S. lycopersicum* Moneymaker	ND	–	Guo et al. (2016)
	TYLCV	*S. lycopersicum Mi-1.2*	ND	+	Guo et al. (2016)
Light	CMV	*A. thaliana*	–	–	Montes and Pagán (2019)
	Oilseed rape mosaic virus	*Arabidopsis thaliana*	+	ND	Yao et al. (2012)
	Tobacco rattle virus	*N. benthamiana*	–	–	Kotakis et al. (2011)
	Tomato mosaic virus	*S. lycopersicum*	ND	-	Matsuura and Ishikura (2014)
	TuMV UK 1	*A. thaliana*	+	–	Montes and Pagán (2019)
	TuMV JPN 1	*A. thaliana*	–	–	Montes and Pagán (2019)
Drought	*Cauliflower mosaic virus*	*A. thaliana*	o/+ (– in 13% genotypes)	+	Bergès et al. (2021)
	Grapevine fanleaf virus	*Vitis vinifera*		–	Jež-Krebelj et al. (2022)
	TuMV	*A. thaliana*	ND	+	Manacorda et al. (2021)

	TuMV	*A. thaliana* Less susceptible	ND	+	González et al. (2021)
	TuMV	*A. thaliana* More susceptible	ND	–	González et al. (2021)
Temperature + CO_2	CMV	*N. benthamiana*	○	○	Del Toro et al. (2017)
	Potato leafroll virus	*Solanum tuberosum*	○	ND	Chung et al. (2017)
	PPV	*N. benthamiana*	–	–	Aguilar et al. (2020)
	PPV	*A. thaliana*	–	–	Aguilar et al. (2020)
	PVX	*N. benthamiana*	–	–	Del Toro et al. (2017)
	PVY	*N. benthamiana*	–	– (○ for 36% isolates)	Del Toro et al. (2017); Makki et al. (2021)
Temperature + Drought	TuMV	*A. thaliana*	+	+	Prasch and Sonnewald (2013)
Temperature + CO_2 + Drought	PPV	*N. benthamiana*	–	+	Aguilar et al., 2020
	PPV	*A. thaliana*	–	+	Aguilar et al. (2020)

(+) indicates increased values under stress; (–) indicates reduced values under stress; (○) indicates unaltered values under stress; and ND indicates that the effect was not determined.

3. Climate change and plant virus transmission

Transmission is arguably the most important trait for plant viruses: In spite of enhanced infectivity, higher virulence and/or optimal levels of within-host multiplication, the virus population will go extinct if successful between-host transmission is not achieved (Elena et al., 2011, 2014). Owing to its importance, the effect of environmental conditions on plant virus transmission (particularly through vectors) is perhaps the most analyzed aspect in the context of climate change (Fig. 4).

3.1 Horizontal transmission

Plant viruses are horizontally transmitted from-plant-to-plant mostly by vectors (Fereres and Raccah, 2015; Lefeuvre et al., 2019), aphids (30% of plant viruses) and whiteflies (20%) being the most important ones (Brault et al., 2010; Hogenhout et al., 2008). Indeed, together these two groups transmit more than 500 species of plant viruses (Fereres and Raccah, 2015). Plant viruses can be transmitted through these insect vectors in several ways: (i) nonpersistently: the virus is retained in the vector stylet, and the

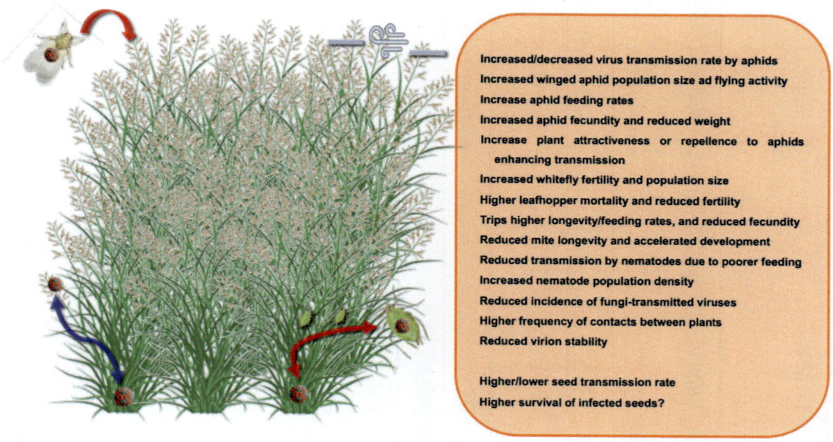

Fig. 4 Schematic representation (left) and summary (right) of reported effects of climate change on plant virus horizontal and vertical transmission. Scheme shows effects on plant attractiveness to vectors (upper left); effects on transmission by contact (upper right); effects on seed transmission (lower left); effects on soil-borne viruses (lower middle); and effects on vector population size and transmission efficiency (lower right). Summary is divided in two blocks: top, effects on horizontal transmission; bottom, effects on vertical transmission. Scheme was constructed using BioRender.

insect remains viruliferous for seconds to minutes after virus acquisition; (ii) semipersistently: viruses are internalized in the insect gut, but do not invade vector tissues resulting in an infectious period of hours to days; and (iii) persistently: viruses are taken up into and retained by insect tissues and invade the salivary glands, with a viruliferous period of days to weeks (Dietzgen et al., 2016; Hogenhout et al., 2008). Other plant virus vectors include leafhoppers, thrips, nematodes and fungi (Hull, 2014). Some of these, like the Western flower thrip (*Frankliniella occidentalis*) are considered as supervectors and mediators of plant virus emergence (e.g., tospovirus and Ilarvirus). Many vectors are hosts for the viruses, and virus replication has been shown to contribute to this enhanced transmission capacity by, for instance, increasing feeding rates (Gilbertson et al., 2015). According to the Disease Triangle paradigm, vectors are central players in the epidemiology of plant virus diseases, and their population dynamics as well as their interaction with hosts and viruses can be modulated by the environment. In addition, some damaging plant viruses, such as tobamoviruses and some potexviruses, can be horizontally transmitted by direct contact through wounds generated by leaf-to-leaf rubbing, soil abrasions during root growth, or mechanically by lawn and farm equipment, grazing animals, pollinators, etc. (Roberts, 2014; Shipp et al., 2008). In this section, we summarize work on the effect of climate change in the efficiency of the different ways of horizontal transmission.

3.1.1 Aphids

Aphids are the most diverse vectors in the temperate zone of the Northern Hemisphere. Aphids exhibit polyphenism, such that genetically identical individuals can potentially show different phenotypes, for instance, they can be apterous or winged (Braendle et al., 2006), the latter being the main drivers of plant virus transmission. A considerable number of studies have addressed the environmental conditions that affect the production of winged individuals in aphid populations: crowding, interspecific interactions, host plant quality and abiotic factors induce winged morph production in aphids (Brisson, 2010; Zhou et al., 1995). All these traits are susceptible to be modified under climate change (Brisson, 2010), thus affecting virus transmission. In addition, environmental cues may also directly modulate aphid transmission by altering the efficiency of acquisition/release of virions, vector behavior, etc.

There is a considerable number of experimental analyses of the effect of conditions linked to climate change on aphid species that have an important impact on plant virus transmission (reviewed by Canto et al., 2009; Trębicki, 2020; van Munster, 2020). As a representation of the work done on this subject, field experiments showed that higher temperature increases the winged population size and flying activity of the lettuce aphid (*Nasonovia ribisnigri*) (Diaz and Fereres, 2005), which positively affects plant virus transmission in lettuce and broccoli (Nebreda et al., 2004). Similarly, higher temperature increased the proportion of BYDV viruliferous aphids in cereal fields (Fabre et al., 2005). Also, eCO_2 may increase aphid feeding rates (Sun et al., 2016), which favors virus transmission (Fereres and Moreno, 2009; Wamonje et al., 2020). A severe water deficit applied to *Brassica rapa* and *A. thaliana* CaMV- or TuMV-infected plants enhanced aphid transmission without altering within-host viral accumulation (Bergès et al., 2018, 2021; van Munster et al., 2017). However, this effect seems to be virus specific as *Turnip yellows virus* (TuYV) aphid transmission was lower in *A. thaliana* plants under water deficit (Yvon et al., 2017), and similar results were obtained for soybean plants infected by *Soybean mosaic virus* (Nachappa et al., 2016). This suggests the action of complex processes that do not depend exclusively on the vector, the plant or the virus, but on double or triple interactions. Indeed, climate change effects can be mediated by the plant infection status. For instance, combined eCO_2 and temperature increased noninfected wheat growth, biomass, and carbon to nitrogen (C:N) ratio, which in turn significantly decreased bird cherry-oat aphid (*Rhopalosiphum padi*) fecundity (Moreno-Delafuente et al., 2020; Trębicki et al., 2016). Similar results were obtained in melon plants colonized by *Aphis gossypii* (Moreno-Delafuente et al., 2021). In contrast, BYDV infection reduced chlorophyll content, biomass, wheat growth and C:N ratio, resulting in a significant increase of *R. padi* fecundity (Moreno-Delafuente et al., 2020). Hence, virus presence alters plant chemistry such that vector population size is enlarged, which enhances the chances for virus transmission. In agreement, a 4-year study of BYDV prevalence under eCO_2 indicated an increasing number of infected plants over time (Trebicki et al., 2017). Higher PLRV transmission by *Myzus persicae* in potato under combined eCO_2 and higher temperature also has been reported (Chung et al., 2017). Experiments with *M. persicae* and bell pepper reported equivalent positive effects of eCO_2 in plant chemistry and vector phenology, but analyses of CMV transmission rate indicated a twofold decrease (Dáder et al., 2016). A possible explanation for this negative correlation between aphid population size and transmission rate was provided by Vassiliadis et al. (2018), who showed that eCO_2 favored the growth of nonviruliferous aphids and decreased the weight of viruliferous ones.

How environment-mediated changes in virus transmissibility are controlled at the molecular level is still poorly understood (van Munster, 2020). Many studies have provided evidence that viruses might alter specific aspects of host plant phenotypes and/or vector behavior to enhance their transmission (Mauck et al., 2018, 2019). For instance, plant viruses are thought to manipulate the amino acid composition and viscosity of plant phloem (Mauck et al., 2010, 2014); downregulate plant defense pathways against vectors though modifications of hormone signaling (Guo et al., 2017; Wu et al., 2017); alter composition of volatiles emitted by the plant (Claudel et al., 2018; Eigenbrode et al., 2002), and enhance leaf yellow color (Fereres et al., 1999; Salvaudon et al., 2013), such that infection increases plant attractiveness or repellence to the vectors to enhance transmission (Carmo-Sousa et al., 2014; Li et al., 2014). It has been shown that these strategies depend on the mode of transmission: viruses transmitted in a persistent manner attract and arrest insect vectors, favoring settlement, reproduction and colony formation in the infected plant, as these viruses require long feeding times for transmission (Mauck et al., 2012). On the other hand, nonpersistently transmitted viruses have no effect, or induce a pull-and-push effect, for vectors because their acquisition is rapid and the time the vector remains viruliferous is short (Bosque-Perez and Eigenbrode, 2011; Carmo-Sousa et al., 2014; Mauck et al., 2012). Although how these manipulative strategies are modulated by environmental changes remains largely unexplored, it is intriguing that the effects of combined eCO_2 and temperature on virus transmission appear to be positive for persistently transmitted viruses (BYDV or PLRV) and negative for nonpersistently transmitted ones (CMV). The few experimental analyses on the molecular basis of this interaction seem to support this transmission mode-specific effect. For instance, eCO_2 activates hormone-mediated defense signaling pathways that alter aphid feeding such that longer periods would be needed for virus acquisition (Guo et al., 2017), which would negatively affect the transmission of nonpersistent viruses; and the reduced transmission of CMV has been linked to poorer *M. persicae* salivation. In contrast, BYDV invasion of *R. padi* activates the expression of heat shock proteins, leading to thermotolerance (Porras et al., 2020). Notably, the opposite situation can be derived from analyses on the effect of drought on aphid transmission (CaMV and TuMV vs TuYV, see above). Obviously, the observed differences may be just due to variations according to the host genotype, virus isolate, and vector biotype involved (Trębicki et al., 2015), and more work is needed to support or reject these apparent trends.

3.1.2 Whiteflies

Whiteflies are both important pests (Aregbesola et al., 2019) and effective vectors for many economically important viruses (Mafongoya et al., 2019). Much of the work on the consequences of climate change for whitefly-transmitted viruses focused on *Bemisia tabaci*. This whitefly species adapts easily to new host plants and geographical regions and, from its original geographic distribution in tropical and subtropical areas, has now been expanded to all continents except Antarctica. It also has ample host range (more than 600 plant species), and transmits viruses from seven different families, including begomoviruses that are the largest group of known plant viruses (Oliveira et al., 2001). This combination of ample geographical distribution (in part associated with climate change, Kriticos et al., 2020) and host range, and high capacity to transmit plant viruses explains research interest. In general, *B. tabaci* populations are favored by high temperatures and moderate rainfall (Sseruwagi et al., 2004), and elevated eCO_2 also increases (Peñalver-Cruz et al., 2020) or has no effect (Cumutte et al., 2014) on whitefly fertility. These environmental conditions, expected with climate change, would enlarge *B. tabaci* population size and therefore would promote virus transmission. Additionally, Peñalver-Cruz et al. (2020) and Roy et al. (2021) looked at the transmission of the begomovirus TYLCV by *B. tabaci* exposing source and receptor tomato plants to ambient or eCO_2 levels and to different temperatures, and observed that *B. tabaci* transmitted the virus at the same rate regardless the CO_2 levels and better at 25 °C than at 35 °C. Together, studies above show that climate change conditions may result in higher vector population size without loss of transmission efficiency, which is likely to enhance virus prevalence. In agreement, using a process-oriented climatic niche model, Kriticos et al. (2020) associated climate change with higher abundance of *B. tabaci* and of cassava virus diseases in East Africa, and validated the conclusion of the model with empirical data. A relationship between higher temperature, whitefly population size and *Okra yellow vein mosaic virus* has been also reported (Ali et al., 2005). In the same line, silencing of whitefly heat shock proteins reduced both its survival rate and ability to transmit TYLCV (Kanakala et al., 2019). Optimal growth rates at higher temperatures (24 °C) also have been predicted for other whitefly species such as *Trialeurodes vaporariorum* (Gamarra et al., 2020b). However, at this temperature transmission rate of *Potato yellow vein virus* was much lower (10%) than at 15 °C (70%). Using this data, (Gamarra et al., 2020a) developed a model of virus infection risk, which predicted that, under climate change, disease risk will be reduced in warmer tropical areas and increased in temperate ones (Gamarra et al., 2020a). These studies indicate that, at least for this vector, the main driver of infection risk is the effect of climate change on the efficiency of virus transmission, rather than on vector population size or a combination of both.

3.1.3 Other vectors

Other arthropods (leafhoppers, beatless, trips, mites) and other organisms (nematodes, fungi) are also virus vectors, some with great socioeconomic impacts (Gilbertson et al., 2015). Studies on the potential effect of environmental conditions on the viral diseases mediated by these vectors are much less abundant than for aphids and whiteflies (Jones, 2016; Trębicki, 2020), but still meaningful to understand the consequences of climate change on plant virus epidemics.

Leafhoppers are the third most important group of hemipteran vectors after aphids and whiteflies, transmitting 4% of all known plant viruses (Hogenhout et al., 2008). Drought does not seem to affect leafhopper abundance (Masters et al., 1998), although it has been reported that higher temperature may reduce the fertility of the leafhopper vector *Dalbulus maidis* (Van Nieuwenhove et al., 2016). Analyses with *Psammotettix alienus*, which transmits *Wheat dwarf virus* (WDV), also reported higher mortality rates at 35 °C than at 25 °C. Interestingly, at the lower temperature WDV transmission rate was highest (Parizipour et al., 2018), suggesting that global warming may negatively affect not only leafhopper population size but also the transmission capacity of the vector.

Thrips are also vectors of highly damaging viruses such as tospoviruses (Rotenberg et al., 2015). Variation in tospovirus transmission rates is observed within and among thrips and tospovirus species (Okuda et al., 2013), and affected by the environmental conditions. For instance, melon thrip (*Thrips palmi*) longevity was higher at 31 °C than at 16 °C, but fecundity was higher at warmer temperatures (Yadav and Chang, 2014). Perhaps the most studied interaction involving a thrip includes *F. occidentalis* and *Tomato spotted wilt virus* (TSWV), a highly damaging tospovirus. Liu et al. (2014) showed that *F. occidentalis* can adapt to eCO_2 by increasing the activities of two types of detoxifying enzymes: carboxylesterase and microsomal mixed-function oxidases and by decreasing the activity of protective enzymes (superoxide dismutase). Remarkably, eCO_2 does not affect thrip population size, but increases feeding rates (Heagle, 2003), which may promote virus transmission. Analyses of temperature dependence of *F. occidentalis* population dynamics indicated negative effects above

26–27 °C (Fatnassi et al., 2015; Li et al., 2014). Exposure to heat for short periods of time also resulted in reduced vector lifespan and fecundity (Jiang et al., 2014; Wang et al., 2014b). These works indicate that *F. occidentalis* can thrive at relatively warm temperatures, but not under heat stress. Interestingly, maximum TSWV transmission is achieved at 29–30 °C, which seems to be counteradaptive for the virus. However, the same study showed that TSWV can modify thrip thermotolerance such that infected thrips had higher survival than noninfected ones as temperature increases (Stumpf and Kennedy, 2007). Heat has been shown to activate the expression of heat shock proteins in the thrip (Wang et al., 2014b). Perhaps virus-induced thermotolerance can be achieved by a similar mechanism. Hence, conditions associated with climate change may increase virus prevalence. Indeed, using a multivariate modeling approach, Chappell et al. (2013) predicted higher TWSV incidence associated with warmer winters that allowed higher vector numbers from the previous summer to survive during unfavorable seasons, which fit with their field observations.

As for mite vectors, Miller et al. (2015) reported that populations of *Aceria tosichella* (wheat curl mite, WCM), vector of WSMV, were unaffected by eCO_2. At odds, Wosula et al. (2015) examined the effects of temperature on WCM survival, which decreased 10-fold from 10 to 30 °C. However, in the same range of temperatures Karpicka-Ignatowska et al. (2021) observed that rising temperatures accelerated mite development and built a model using this data predicting that the risk of WCM was highest at 25–30 °C. Interestingly, field experiments showed that, within the range of optimal temperatures for mite growth, warmer conditions (mean daily temperatures >10 °C) resulted in high rates of mite infestation, leading to higher WSMV prevalence of up to 88% (Ranabhat et al., 2018). Hence, rising temperatures are predicted to increase epidemics of WSMV as global warming progresses, a prediction that could be extrapolated to other mite-transmitted viruses all else being equal (Singh et al., 2018).

Plant-parasitic nematodes transmit tobraviruses and nepoviruses (MacFarlane and Robinson, 2004), with rates affected by environmental conditions (Singh et al., 2020). For instance, the efficiency of transmission of *Raspberry ringspot virus* to cucumber seedlings by its nematode vector *Longidorus macrosoma* was lower as temperature increased, until no virus transmission was detected at 30 °C. This effect seemed to be associated with poorer nematode feeding rather than changes in vector survival (Debrot, 1964). Although in this work survival was unaffected by temperature, environmental conditions may have a significant impact on nematode reproductive success, modulating population sizes and therefore chances for virus transmission. For instance, reproduction of chickpea nematodes surveyed in Spain increased with rising temperatures (Castillo et al., 1996). Similarly, Neilson and Boag (1996) identified temperature as a major driver of longidorid and trichodorid growth in the UK, which transmit nepoviruses (Singh et al., 2020). The population density of the nematode vector *Longidorus elongatus* in grass-dominated pasture was also increased by eCO_2 (Yeates and Newton, 2009). Because nematode vectors require host plants to complete their life cycle, climate change effect on host plants are likely to affect chances for virus transmission. For example, benefits of eCO_2 can be compensated by changes in plant chemistry and defenses that have negative effects in the nematode population density (Cesarz et al., 2015). Again, this scant work draws a picture of a complex interaction between the virus, the host, the vector and the environment.

As all organisms mentioned above, fungal vectors are also subjected to changes in their population dynamics due to environmental conditions. For instance, *Olpidium* species, which transmit agronomically important viruses such as *Pepino mosaic virus* (PepMV) or *Melon necrotic spot virus* (MNSV), optimally grow at warmer temperatures (around 20 °C) (Gharbi and Verhoyen, 1994). Environmental conditions associated with climate change may also modulate virus transmission through fungi: Increased soil moisture and low air temperature promote activity and movement of vector zoospores and therefore the fungal vector capacity (Kühne, 2009). In these conditions, occurrence of fungus-transmitted *Lettuce big-vein virus* has been shown to increase (Campbell, 1996). Consequently, climate change conditions leading to less frequent rainfalls, longer drought periods and/or desertification would reduce the incidence of fungi-transmitted viruses (Jones and Barbetti, 2012). In a previous review, Jones (2016) highlighted a lack of published information on the influences of eCO_2 on fungal virus vectors, a gap that has not been covered at the time of this review.

3.1.4 Transmission by contact

Contact-transmitted viruses are probably the ones more exposed to environmental conditions as many of them have stable virions that allow them to persist outside of the host (Hull, 2014). Then, climate change conditions such as altered temperature, soil moisture, heavy rainfall and windstorms are likely to have an effect in their survival, virulence and chances for transmission. However, little work has been done involving this type of virus. Heavy rainfall would create water splashing, favoring the dispersion of viruses contaminating the soil (Jones, 2016). In addition, Gerrard (2013) showed that exposure to both flooding and infection with the contact-transmitted *White clover mosaic virus* (WClMV) on white clover resulted in higher yield losses as compared with effect of any of the two stresses separately. Contact-transmitted PepMV survived for weeks in the water, which was a source of inoculum for these viruses in tomato plants (Mehle et al., 2014), but wet soils do not increase TMV survival, another contact-transmitted virus (Fraile et al., 2014). It also has been reported that higher numbers of contacts between a TMV-infected plant and a noninfected susceptible one increases virus transmission rate (Sacristán et al., 2011). Hence, higher frequency of contacts between plants, as those occurring from strong winds, would favor virus transmission. At odds with these potential positive or neutral effects of climate change on transmission by contact, higher temperature may reduce virion stability (Chakravarty et al., 2020), challenging the survival of viruses outside the host (Orlov et al., 1998). A similar rationale can be applied for mechanical virus transmission.

3.2 Vertical transmission

Horizontal transmission is not the only way for virus dispersal. About 25% of all known plant viruses are vertically transmitted from parents to offspring through the seeds, seed transmission being a major component for the fitness of these viruses (Pagán, 2019; Simmons and Munkvold, 2014). Virus seed transmission may have a high impact in plant virus epidemiology (Pagán, 2022; Sastry, 2013) as it: (i) provides the virus with a means to persist for long periods of time when hosts or vectors are not available, as many seed-transmitted viruses can survive within the seed as long as it remains viable (Sastry, 2013). (ii) Allows also for long distance dissemination of the virus via infected seeds (Albrechtsen, 2006). Indeed, evidence exists that bird dispersion and human trade of infected seeds allowed cross-continental jumps of some plant viruses (Dwyer et al., 2007; Vincent et al., 2014). (iii) Represents an important source of primary inoculum for many seed-transmitted viruses, which are disseminated afterward via insect vectors. Therefore, even extremely low seed transmission rates can initiate damaging epidemics (Maule and Wang, 1996; Sastry, 2013). For instance, in Europe the acceptable threshold limit of lettuce seeds infected by *Lettuce mosaic virus* was 0.1%, until it was demonstrated that a percentage of infected seeds as low as 0.003% was enough to start an outbreak (Albrechtsen, 2006; Grogan, 1980). This is particularly important as many vectors transmit viruses in a nonpersistent manner (see above), which means that insecticides are not effective at suppressing virus spread (Sastry, 2013), and seed transmission may greatly contribute to increased virus prevalence.

Owing to these benefits, seed transmission may confer crucial advantages for viruses in a climate change scenario allowing survival in conditions, such as prolonged drought periods or extreme weather events, when hosts and/or vectors are not available; and facilitating virus geographic range expansion when climatic conditions become favorable in distant regions that viruliferous vectors would not reach (Jones, 2016; Pagán, 2022). The few existing experimental analyses on the effect of the environment on seed transmission rarely consider more than one climatic variable at the same time. For instance, seed transmission is generally reduced at higher temperatures and under drought conditions (Coutts et al., 2009; Johansen et al., 1994; Jones and Proudlove, 1991) and is favored by higher light intensity (Montes and Pagán, 2019). However, there is no report on the effect of eCO_2 on virus seed transmission and interactions between climatic variables have not been analyzed. The exception comes from a work dealing with a particular type of vertical transmission: infection of daughter potato plants through tubers. Bertschinger et al. (2017) reported higher PVX, PVY and PLRV tuber transmission rates in locations with higher diurnal temperature and low light intensity and rainfall regime, with the latter environmental factor chiefly explaining virus vertical transmission. Interestingly, these studies focused on the effects of environmental conditions of seed transmission rate (percentage of infected seeds). Although relevant, this measure of the efficiency of seed transmission does not account for the effect of virus infection on plant progeny production. Indeed, it has been proposed that the total number of infected progeny reflects more accurately the contribution of vertical transmission to pathogen fitness and therefore is more directly linked to its evolution (Anderson and May, 1982; Hamelin et al., 2016; Lipsitch et al., 1996). For instance, high seed transmission rates at the cost of extreme reduction in seed production would be evolutionarily less advantageous for the virus than intermediate levels of seed transmission and virulence. Hence, the reports on the effect of environmental variables on seed transmission rates provide only partial information on the potential consequences of the impact climate change for virus epidemiology.

Additional information can be extracted from analyses dealing with the effect of climate on the factors that control seed transmission. For seed transmission to occur, it is crucial that the virus reaches and invades plant reproductive organs before gametogenesis and/or while the embryo is still accessible from mother cells, without affecting gamete/embryo viability

(Bradamante et al., 2021; Cobos et al., 2019; Maule and Wang, 1996). Thus, it has been proposed that the number of infected progeny would be determined by: (i) the ability of the virus to reach and invade gametic tissues, which is linked to virus within-host multiplication and movement; (ii) the plant progeny production upon infection, associated with virus virulence and plant tolerance; and (iii) the embryo survival in the presence of the virus (Hamelin et al., 2016; Lipsitch et al., 1996; Maule and Wang, 1996). Experimental analyses provided support for these predictions (Cobos et al., 2019; Pagán et al., 2014). As discussed in Section 2, most of these determinants of seed transmission are affected by environmental conditions associated with climate change such as drought, eCO_2, raised temperatures or higher light intensity. Generally, reported results indicate that these environmental conditions reduce virus multiplication and virulence, which would favor seed transmission (Pagán et al., 2014). However, the interaction between climatic conditions and virus infection determines seed survival, and the potential consequences for seed transmission remain to be elucidated. It has been shown that virus presence may increase the survival of aged seeds (Bueso et al., 2017), where seed aging was done through a modification of a thermotolerance assay. Thus, results indirectly suggest that viruses may confer seeds with tolerance to raising temperatures, which would be beneficial under climate change, and may enhance the importance of seed transmission for virus epidemics.

Not every vertically transmitted virus needs to actively invade seeds or tubers. Persistent (also known as cryptic) viruses (*Partitiviridae, Endornaviridae, Chrysoviridae, Totiviridae*) do not move systemically, but are found in every cell, including reproductive structures in every plant generation (Boccardo et al., 1987). Partitiviruses seems to be highly abundant in wild species (up to 60% of the viral diversity) and undergo strict vertical transmission via meiosis (Roossinck, 2012a). Because these viruses generally cause asymptomatic infections, it has been proposed that they would be beneficial for their hosts at least under certain conditions (Roossinck, 2015). Some evidence supports this idea. For instance, *White clover cryptic virus* prevents the formation of nitrogen-fixing nodules in white clover if there is enough nitrogen in the soil (Nakatsukasa-Akune et al., 2005). Hence, the benefits they confer could be crucial for host survival under global warming or at eCO_2 that alters plant C:N balance. There is a lack of work on how environmental conditions affect the transmissibility of persistent viruses, but jalapeño pepper plants infected by *Pepper cryptic virus 1* (*Partitiviridae*) deterred *M. persicae*, a common vector of acute viruses.

This observation points again to a positive effect of persistent viruses on plants, here by protecting them from the infection by other viruses and from herbivory (Safari et al., 2019).

4. Climate change and plant virus ecology

Besides altering the rate of plant virus transmission, environmental conditions may impact other infection traits that are relevant for virus epidemiology (Fig. 5). Perhaps one of the most important relates to the ecological relationships established by plant viruses, particularly the interaction of viruses with the host at the community level (host range), and with other pathogens (mixed infections).

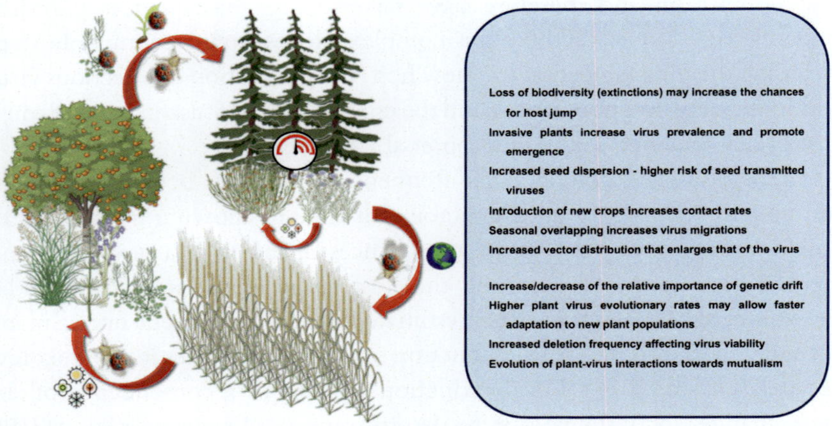

Fig. 5 Schematic representation (left) and summary (right) of reported effects of climate change on virus ecology and evolution. Scheme shows: new contacts between wild populations due to plant invasion, dispersion of infected seeds and geographical expansion of vectors (upper left); new contacts due to virus movement from wild to cultivated plant populations due to geographical expansion of vectors and introduction of crops in new geographic areas (right); new contacts due to virus movement from cultivated to wild plant populations due to geographical expansion of vectors and seasonal overlapping (bottom left). Small red arrow indicates new contacts within the same plant community due to seasonal overlapping of different susceptible species. Speedometer indicates faster virus evolution promoting adaptation. Summary is divided in two blocks: top, effects on plant virus host range/emergence; bottom, effects on plant virus evolution. Scheme was constructed using BioRender.

4.1 Host and geographic range

A key trait determining virus relationships with the plant community is its host range, which is determined by the distribution, abundance and interaction of host species (McLeish et al., 2018). Viruses can be generalists that have broad host ranges but are often poorly adapted to individual hosts, or specialists which have narrow host ranges but are well adapted to their hosts (Futuyama and Moreno, 1988; Whitlock, 1996). Generalists tend to be better adapted than specialists to alterations in host range under climate change scenarios and are expected to become more prevalent (Jones, 2016). Generalists host range is dynamic and can be regulated either by: (i) the availability of susceptible hosts (local host range); it has been shown that the ecosystem composition determines the host range of certain viruses such that, in extreme cases, generalist may behave as specialists (McLeish et al., 2018; Valverde et al., 2020); or (ii) by processes of host range expansion (host jumps). Host jumps underpin numerous cases of emergence (i.e., the increase of disease incidence following virus appearance in a new, or previously existing, host population) and therefore have far-reaching consequences for virus epidemiology (Jones, 2009). This complex process involves multiple steps including the initial infection in a new host, the production of infectious virus within the new host population, and the establishment of sustainable transmission cycles (Holmes, 2009; Hudson et al., 2008).

Ecological factors, such as the frequency of contacts between reservoir and novel hosts, are of fundamental importance in allowing initial steps of host jumping as they determine the chances of infecting a new host, and therefore are central in defining the host range (Anderson et al., 2004; Keesing et al., 2010). It has been estimated that, at its current rate, climate change will lead to the global extinction of one in six species in the next few decades (Urban, 2015). Local extinction can also be a consequence of climate change, for instance due to desertification (Mirzabaev et al., 2019). Depending on whether plant species prone to extinction are better or worse virus reservoirs, climate change may result in higher or lower virus prevalence, and hence may or may not favor host jumps. Interestingly, it has been reported that more competent hosts are more resistant to extinction (Joseph et al., 2013). Accordingly, it could be predicted that loss of ecosystem biodiversity due to climate change would enhance virus prevalence and the chances for a host jump by increasing contact rates between susceptible plant species, which would be more abundant.

Climate change may also alter contact rates by the opposite phenomenon, that is, by increasing plant diversity. As regions in formerly cooler higher latitudes gradually become warmer, and when rainfall remains sufficient, conditions may become favorable for plant species not previously present in these regions (Jones, 2016). Additionally, dust storms favor the dispersion of plant seeds, which can promote invasion events (Womack et al., 2010). Hence, climate change is predicted to significantly change vegetation communities around the world and to promote plant invasions (Dai et al., 2022). This would have a major impact on contact rates and therefore on plant virus prevalence. Depending on the nature of the invasive species, two hypotheses, developed to understand the effect of ecosystem biodiversity on disease risk (Keesing et al., 2010; Ostfeld and Keesing, 2012), may predict the outcome of plant invasion. The "Amplification Effect" hypothesis predicts that higher biodiversity increases disease risk. Thus, it assumes that new (here, invasive) species will be competent host for viruses, increasing contact rates and therefore resulting in higher abundance of inoculum sources and virus prevalence. The "Dilution Effect" hypothesis predicts a negative correlation between biodiversity and disease risk. In this case, invasive species will be less susceptible to infection or nonhosts, therefore reducing contact rates and virus prevalence. In general, invasive species can be more robust in a new environment because they have left behind pathogens in their native habitat, a phenomenon known as pathogen release (Mitchell and Power, 2003; Roossinck, 2013). Indeed, Mitchell and Power (2003) found that, on average, 24% fewer virus species infect each plant species in its naturalized vs in its native range. Hence, invasion would potentially result in lower rather than in higher prevalence. However, several examples indicate the opposite. For instance, plant species can be aided in their invasions by plant viruses in various ways: They may carry viruses in asymptomatic infections that cause disease in their native competitors, or they may be more tolerant than natives to viruses that are already in the colonized environment (Rúa et al., 2011). Although a recent study would argue against this idea, as it shows that the invasive species *Eugenia uniflora* was more negatively affected by soil microbiota (fungi, bacteria and viruses) than native *Eugenia* species, prevalence of each type of microorganism was not analyzed and the work did not consider other nonsoil-borne viruses (von Holle et al., 2020). Invasive plants may also play a role in the ecology of plant virus vectors. One study showed that an invasive grass species resulted in increased populations of the aphid vector for plant viruses that was detrimental to a native grass species (Malmstrom et al., 2005). Similarly, Szabó et al. (2020) showed that higher abundance of an invasive weed increased both vector abundance and virus prevalence in nearby crops. Finally, invasive species may carry beneficial viruses that give them a competitive advantage in comparison to native species. We will discuss reports on tolerance to drought, temperature and elevated CO_2 conferred by certain acute virus infections in Section 5, and we mentioned above that persistent viruses can also provide such benefits. If this type of infection outcome is more frequent in invasive than in native plants in the naturalized habitat, these species would acquire a selective advantage under climate change. Introduction of new plant species cannot only occur through invasive species, but also by movement to a new location of food crops domesticated elsewhere as climate change makes the environment suitable for growing them. In this new ecological context, contact rates will be altered such that crops will be challenged by damaging virus diseases that they never encountered before, leading to disease emergence. Recently, this subject was thoroughly reviewed (Jones, 2020).

Contact rates also can be altered even if plant community composition remains unchanged. For instance, global warming will allow expanding the growing period for certain crops (Jones, 2009; Jones and Barbetti, 2012). Consequently, plants in adjacent wild ecosystems or weeds, which were previously seasonally isolated (particularly, annual plants), would become sources of inoculum for the expanded crops and vice versa. This will have a major impact in agriculture as the accompanying flora is a main source of primary inoculum for crop infection and allows virus overwintering (Barreto et al., 2013; García-Arenal and Zerbini, 2019). Moreover, this phenomenon may threaten the preservation of wild ecosystem as viruses present in crops may have major impacts on the fitness of wild plants. For instance, in North America cereals act as sources of BYDV inoculum for *Panicum virgatum* (switchgrass), a native prairie grass, whose infection results in a fitness reduction of 30% (Alexander et al., 2017). Similar climate change-induced seasonal overlapping between crops or species within the same wild ecosystem would also increase contact rates, virus host ranges and potentially virus emergence (Valverde et al., 2020). A similar rationale can be applied for parasitoids: Changes in temperature may disrupt the synchrony between vector and parasitoid life cycles (Hance et al., 2007), resulting in a poorer control of the vector population and therefore in higher chances for virus transmission.

Changes in vector distribution may also increase virus host and geographic range in the absence of changes in the plant community composition. Climate change creates new ecological niches that provide opportunities for vectors to establish and spread in new geographic regions and shift from one region to another (Islam et al., 2020). For instance, it has been shown that BYDV significantly enhances the thermal tolerance of its aphid vector *R. padi.*

This enhanced thermal tolerance allowed aphids to expand their fundamental niche, and therefore that of the transmitted viruses, reaching plant communities in latitudes that were previously virus free (Porras et al., 2020). This process of expansion of the vector geographic range has been repeatedly invoked to explain disease emergence due to climate change and a number of examples can be found in the literature (e.g., Canto et al., 2009; Jones, 2016; Jones and Naidu, 2019 and references therein).

Although contact rates are central to understand plant virus host range and disease emergence, many parasites often replicate poorly in new hosts and hence are inefficiently transmitted, such that host, and virus genetic factors also need to be invoked to explain successful host range expansion (Pepin et al., 2010; Elena et al., 2014). On the host side, the type and strength of host defenses is a major barrier that viruses may overcome to achieve a successful host jump. On the virus side, standing virus population genetic diversity in reservoir hosts facilitates infection in new hosts as it increases the chances that preadapted variants are present in the virus population (Elena et al., 2014; Mattenberger et al., 2021). Also, high evolutionary rates of plant viruses facilitate the process of adaptation to new hosts (Elena and Sanjuán, 2005; Holmes, 2009). Consequences of environmental variation for the rate of virus evolution and the genetic diversity of virus populations will be addressed in Section 5, and the effects of climate change on host defenses will be discussed in Section 6 as part of control measures.

4.2 Mixed infections

Plant viruses not only establish relationships with their host plants or with their vectors, but also with other viruses. Because a large fraction of known plant viruses are multi-host pathogens (García-Arenal and Fraile, 2013; Moury et al., 2017; Roossinck, 2012b), it is likely that more than one virus infects the same plant (mixed infections). Although initially considered as relatively infrequent, increasing evidence indicates that mixed infections are common in nature (Roossinck et al., 2010; Syller, 2012; Tugume et al., 2016). The importance of mixed infections in the epidemiology and genetic diversity of plant virus populations was realized almost a century ago (Fawcett, 1931). Early works indicated that mixed infections could lead to the appearance of new virus strains and/or to modify the fitness of the cohabiting viruses (Rochow, 1972), and both synergistic (Nsa and Kareem, 2015; Rentería-Canett et al., 2011; Taiwo et al., 2007) and antagonistic (Martín and Elena, 2009; Syller, 2012) effects have been described.

Also, mixed infections may have important consequences for the host as they may enhance virus symptoms (Rodelo-Urrego et al., 2013; Vanitharani et al., 2004), allow breakdown of host genetic resistance (García-Cano et al., 2006), or enhance virus transmission (Hobbs and McLaughlin, 1990; Ryabov et al., 2001). These effects may have a deep impact on the plant community composition (Gómez et al., 2009; Moreno and López-Moya, 2020; Rodelo-Urrego et al., 2015).

As with single infections, virus-virus interactions (as well as those of the two partners with hosts and vectors) are influenced by the environment in which they occur. To date, most work on this topic focused on the effects of temperature. Scheets (1998) showed that infectivity of co-inoculated *Maize chlorotic mottle machlomovirus* and WSMV was higher than that of singly infected viruses at higher temperatures (31 °C), but this advantage disappeared at lower temperatures (26 °C). In this work, higher light intensity enhanced symptom severity in co-infected plants. Mixed infections of PVX and PVY showed milder symptoms at 26–40 °C than at 5–20 °C (Senanayake and Mandal, 2014), and elevated temperature markedly decreased virulence in the PVX/potyvirus-associated synergistic infection. This reduction was associated with lower virus titers, which was the consequence of a significantly enhanced HR-like response elicited by the PVX P25 together with the potyviral helper component-proteinase protein, which are suppressors of the plant RNA silencing defense (Aguilar et al., 2015). In agreement, Alcaide et al. (2021) showed that higher temperature (30 °C) repressed the accumulation of two PepMV strains in mixed infections, but not always when these strains were singly inoculated. The importance of temperature for the outcome of mixed infections recently led to the development of a mathematical model aiming at predicting thermal reaction norms for plant virus mixed infections. This model (parameterized with data derived from infections by two PepMV isolates) predicts that mutual interference of virus strains on replication is stronger at higher than at lower temperatures (Sardanyés et al., 2022), in line with Alcaide et al. (2021). At odds, virulence of PVX/potyvirus-associated synergism was maintained at eCO_2, and this condition suppressed the HR-like response (Aguilar et al., 2015). These contrasting results depending on the abiotic stress are not surprising given that it has been shown that the outcome of mixed infections depends on the virus species, the host genotype and the environmental conditions (Tatineni et al., 2010). Thus, rather than a general rule, the interaction between climatic conditions and virus-virus-host interactions would need to be analyzed in a case-specific way, which makes it difficult to develop general-purpose control measures. These results also point to the need of analyzing the interactions between multiple climate variables to fully understand the effect of climate change.

Viruses cannot only establish mixed infections with other viruses, but also with other microorganisms, and these interactions are potentially affected by environmental conditions. For instance, mixed infections of *Bean yellow mosaic virus* (BYMV) and the fungus *Kabatiella caulivora* were more virulent than single infection by any of the two pathogens. Virulence was modulated by temperature such that it decreased at higher (22.5 °C) than at lower (18 °C) temperatures; this affect is apparently explained just by changes in BYMV multiplication (Guerret et al., 2016). In another study, mixed infections of BYDV and an arbuscular mycorrhizal fungus resulted in higher virus multiplication and fungus growth at higher vs lower CO_2, which also resulted in reduced plant growth (Rúa et al., 2013). Hence, these observations seem to be largely in accordance with responses to temperature and eCO_2 in virus-virus interactions. Alternatively, the effect of environmental conditions on mixed infections of viruses with other organisms may depend on the nature of the interacting partner: pathogen-pathogen (Guerret et al., 2016) or pathogen-mutualist (Rúa et al., 2013). More studies are needed to disentangle the dominating effect, if any.

5. Climate change and plant virus evolution

Most plant viruses have genome sizes ranging between 5 and 15 kb, and one of the defining traits of viruses with small genomes is their high capacity to generate genetic diversity (Holmes, 2009; King et al., 2012). This capacity has been proposed to be one of the main reasons for their biological success, as it has been linked to the continuous appearance of new virus genotypes or species that colonize new areas or previously noninfected host populations (Elena et al., 2014; García-Arenal and McDonald, 2003; Holmes, 2009). The high capacity of plant viruses to generate genetic diversity derives from a combination of factors. The first major factor is their high mutation rates. Most plant viruses have RNA genomes that encode RNA-dependent RNA polymerases, which lack proofreading activity resulting in high error rates (Drake et al., 1998). Interestingly, despite using error-proof polymerases, DNA plant viruses have been shown to have mutation rates in the same range as RNA viruses (Pagán and García-Arenal, 2018). High mutation rates provide viruses the capacity for rapid adaption: The higher the mutation rate, the larger the probability of generating virus genotypes with more fitness in new environments (including new hosts and new climatic conditions). That environmental conditions may affect mutation rates of RNA viruses has been known for a long time. Thus, in a review of the effect of temperature in spontaneous mutation rates, Lindgren (1972) cite a work on TMV showing that mutation rate is higher at 27–32 °C than at 17 °C. In the same line, Alcaide et al. (2021) reported higher number of mutations in the tomato-infecting PepMV populations when plants were grown at 30 °C compared to 20 °C, and regardless of the virus being inoculated in single or in mixed infections. Broader analyses considering all dsDNA viruses led to the same conclusion: smaller genome length (which correlates with faster mutation rates, Duffy et al., 2008), were observed in viruses occurring at higher temperatures (Nifong and Gillooly, 2016). In contrast, lack of variation in virus population genetic diversity with temperature has been reported recently (Aimone et al., 2021). Hence, in general climate change may promote higher levels of genetic diversity, which may increase the adaptive capacity of plant viruses. However, most mutations in RNA plant viruses appear to be highly deleterious or lethal, whereas neutral mutations are much less frequent (Carrasco et al., 2007; Hillung et al., 2015). In this scenario, high mutation rates may lead to a high burden of deleterious mutations (i.e., mutational load) in the virus populations and ultimately to extinction (Chao, 1990). Moreover, lack of polymerase fidelity not only result in higher point mutations, but also in insertions and deletions. It has been shown that higher wheat cultivation temperature increases the frequency of deletions in the genome of the soil-borne *Wheat mosaic virus* RNA 2, which has deleterious effects for virus transmission through its fungal vector (Chen et al., 1995). In this context, the second major factor associated with the high genetic diversity of virus populations becomes important: plant viruses generally have large population sizes (García-Arenal et al., 2001), such that deleterious mutations can be quickly purged from the virus population by negative selection (Elena and Sanjuán, 2005). As discussed in Sections 2 and 3, environmental conditions associated with climate change may promote virus multiplication and transmission, resulting in higher population sizes. Hence, the simultaneous increase of mutation rates and population size may prevent excessive mutational load. However, this is not a general rule and, in certain cases, climate change might potentially lead to virus extinction when higher mutation rates are accompanied by lower virus multiplication/transmission. Finally, the third major factor is the much shorter generation time of RNA viruses (minutes to hours) (Wu et al., 1994) than that of plants (months to years). Thus, virus evolution occurs in different time scales from that of their host plants. This gap might be increased by climate change due to faster virus mutation rates. However, it has been shown that higher temperature also increases plant mutation rates (Belfield et al., 2021). Remarkably, in *A. thaliana* the accumulation of mutations is not uniform across the plant genome, with higher frequency of mutations in defense genes (Lu et al., 2021). Such higher variability might contribute to help the plant keep the pace of pathogen evolution.

Besides mutation, genetic diversity in virus populations can also be generated by recombination, that is, the exchange of genomic fragments between genotypes. Recombination may represent an evolutionary advantage for viruses because (1) it can create fitter genotypes more rapidly than mutation and (2) it might purge deleterious mutations from virus populations, thereby preventing the decrease in overall fitness due to the accumulation of deleterious mutations (Chao, 1990; García-Arenal et al., 2001; Moya et al., 2004). There is little work on the relationship of this aspect of plant virus evolution and climate change, and the few experimental analyses indicated no variation in recombination frequency with temperature (Alcaide et al., 2021). However, this work was focused on PepMV, a virus that in general has low recombination rates, and the situation might be different in other viruses in which this mechanism of generating genetic diversity is more important. Indeed, temperature has been shown to affect recombination rates in animal viruses, but in this case, there was a negative correlation between the traits (Li and Zhang, 2001). In contrast, higher recombination rates as temperature, light intensity and precipitation increase have been widely documented in plant genomes (Dreissig et al., 2019; Modliszewski et al., 2018). These effects may also help to compensate higher plant virus evolutionary rates.

Mutation and recombination are a consequence of the mechanisms of virus replication, and the resulting new genotypes are therefore in theory randomly generated (but see Bujarski, 2013 for exceptions). Central to understanding the population genomics of plant viruses is identifying the determinants of the frequency of the genotypes generated through mutation and recombination in the populations (Duffy et al., 2008). In the absence of migration, the number and frequency of these genotypes in the population (i.e., the genetic structure of the population) is the result of two different evolutionary processes: genetic drift and selection (Acosta-Leal et al., 2011; García-Arenal et al., 2001).

Genetic drift occurs when populations of organisms are not large enough to ensure that each genotype will have progeny in the next generation. Consequently, the genotypes passed into the next generation are randomly sampled from the mother population. Genetic drift may be particularly relevant in plant virus populations during the severe reductions in their population size (population bottlenecks) that may occur in many steps of the virus life cycle (infection of a new host population, a new host plant either through vertical or horizontal transmission, or new organs within a host plant) (Ali et al., 2006; Fabre et al., 2014; Gutiérrez et al., 2012; Li and Roossinck, 2004; Sacristán et al., 2003). Genetic drift reduces the genetic diversity within populations and increases the diversity among populations. Also, because the genotypes that start a new population are not selected according to their fitness, genetic drift counters the effects of selection (García-Arenal et al., 2001; Acosta-Leal et al., 2011). However, genetic bottlenecks may also have positive effects for plant viruses: (i) because not every genotype passes to the next generation, defective genomes may be removed from the virus population, (ii) due to epistatic interactions in virus genomes, fitness landscape are commonly rugged, which may facilitate the virus to move from one fitness peak to another (Zwart and Elena, 2015). Although the size and frequency of genetic bottlenecks in plant virus infections have received considerable attention (Pagán and García-Arenal, 2018; Zwart and Elena, 2015), very little is known on how climate change affect their relative importance for plant virus evolution. We can only hypothesize on the potential effects on virus evolution based on work on traits that may affect such bottlenecks. At the within-host level, higher titers of virus multiplication such as those observed at higher temperature or eCO_2, would reduce the size of the bottleneck during systemic movement (Gutiérrez et al., 2012), decreasing the importance of genetic drift. In contrast, stronger RNA silencing responses against certain viruses under higher temperature (Chellappan et al., 2005) may have the opposite effect. Because within-host virus multiplication is also a major determinant of seed transmission (Cobos et al., 2019), which also represents a population bottleneck (Fabre et al., 2014), the effects of climate change would not only affect within- but also between-host bottlenecks, modulating the importance of genetic drift at different levels. For horizontal transmission, genetic bottlenecks depend on the number of virus particles transmitted by a single vector and on the number of vectors acting as sources of inoculum per plant (Ali et al., 2006; Betancourt et al., 2008; Gadhave et al., 2020). Hence, higher vector population sizes such as those reported under higher temperature and eCO_2 would reduce the impact of genetic drift in the genetic structure of the virus population. At the host population level, longer cropping cycles or improved plant survival may reduce the frequency of extinction-recolonization processes that occur between seasons and/or in annual host plants (García-Arenal et al., 2001), and therefore impact the between-season genetic bottlenecks.

Conversely, geographical expansion of vectors and host plants may lead to virus emergence in new regions through a reduced number of founder individuals, which would also increase the significance of genetic bottlenecks (Weaver et al., 2021).

Selection is a directional process by which the fittest genotypes in a given environment increase their frequency in the population (positive selection), whereas less fit genotypes decrease their frequency (negative or purifying selection). Similar to genetic drift, selection results in a decrease of the population diversity and may also cause an increased diversity between populations, if they are under different selection pressures, so that the effects of selection and genetic drift are often difficult to distinguish. When selection has been differentiated from genetic drift, selection has been associated with every life-history trait of plant viruses, such as survival in the environment due to higher structural stability of the virus particles (Fraile et al., 2014), adaptation to the host plant resulting in more effective within-host multiplication (Hillung et al., 2015), adaptation to host defenses allowing resistance breakdown (García-Arenal and McDonald, 2003), and adaptation to a transmission mode resulting in more efficient between-host dispersal (Pagán et al., 2014). As we have discussed in previous sections, climate change may affect all these virus life-history traits. Hence, environmental conditions will be a major determinant of selection pressures acting on virus evolution. However, experimental evolution studies analyzing this adaptation processes are scant for plant viruses.

Climate change would not only modulate selection on traits related to plant-virus pathogenic interactions. Increasing evidence indicates that virus infections may confer important benefits to their host plants that allow adaptation to climate change such that pathogenic interactions may evolve toward mutualism (Poudel et al., 2021). For instance, it has been shown that some plant viruses confer tolerance to drought (Xu et al., 2008), and that this tolerance is the consequence of virus evolution from parasitism to mutualism (González et al., 2021). This effect has also been described in wood crops like grapevines infected with GFLV, but only under mild water stress as previously mentioned (Jež-Krebelj et al., 2022). Infection by TYLCV confers both thermotolerance and drought resistance to tomato (Aguilar and Lozano-Duran, 2022). A detailed list of examples and of the molecular basis of such beneficial effects can be found in a recent review (González et al., 2020). Similarly, viruses can confer heat tolerance to their vectors (Porras et al., 2020). These benefits may be selectively advantageous for both the virus and the plant under climate change, representing a major selective force. However, these beneficial effects are not always apparent in the presence of more than one abiotic stress. For instance, PVX and PPV infection induced tolerance to drought and resistance to *Pseudomonas syringae* in *N. benthamiana* and *A. thaliana*, but these responses disappeared under combined eCO_2 and temperature (Aguilar et al., 2020). If this is a general rule, and considering that in nature plant viruses are often subjected to multiple and simultaneous abiotic stresses, the potential advantage of being infected would be lower for both the plant and the virus.

6. Climate change and management of plant virus diseases

From the work discussed above it can be concluded that climate change is expected to alter pathosystem dynamics, and consequently the effectiveness of existing strategies to control virus diseases may become unreliable (Jones, 2016). This is not only restricted to alterations in the focal crop, but also to changes in the population dynamics of the accompanying flora (i.e., edge vegetation and weeds) and of adjacent wild ecosystems (Alexander et al., 2014). To cope with this changes, increasingly sophisticated and diverse phytosanitary, cultural, chemical, and biological measures for virus control will be required. In this section, we summarize work on the consequences of climate change for the main disease control strategies: (i) prophylaxis to restrain virus dispersion (quarantine, certification, removal of infected plants, control of natural vectors, or other procedures), and (ii) immunization (genetic resistance obtained by plant breeding, plant transformation, cross-protection, or others).

6.1 Detection and diagnosis

Disease management relies strongly on accurate and efficient identification of the causal agent, which is also central to plant certification. The most utilized techniques for virus identification are based on the serological (ELISA) and molecular (PCR, qPCR, High Throughput Sequencing or HTS) characterization of plant viruses, which must be optimized for accuracy, measured as sensitivity and specificity (Sharma et al., 2009).

Sensitivity measures the proportion of real positives that are classified as such (probability of true positives). Two main factors can be named among the most important for sensitivity: (i) Target accumulation level and (ii) target purification method (Rubio et al., 2020). Several studies shown that environmental conditions associated with climate change reduce virus within-host multiplication (see Section 2). Low virus titer can limit sensitivity, producing false negatives when the virus titer is under the technique detection threshold. In such cases, adaptation of currently available detection methods to the climate change context would require increasing sensitivity to minimize false negatives. Real-time qPCR is becoming the molecular method of choice for routine virus detection as viruses with low accumulation levels can be readily detected. However, the proportion of false positives using this technique can reach 5% (Ahmed et al., 2022). In addition, hybridization and PCR techniques require purification of total RNA or DNA from plants. Climate change greatly alters plant physiology, modifying protein and lipid content, and hormone regulatory networks (Calleja-Cabrera et al., 2020). These alterations may increase the presence of substances inhibiting the detection process. Recently, new methods that allow more efficient purification of viral RNA (Hataya, 2021) or novel techniques, such as digital droplet PCR (Salipante and Jerome, 2020), with detection thresholds well below those of real-time qPCR have been developed. Both also have been shown to reduce the presence of the effects of inhibitors. These technological developments will contribute to the adaptation of detection methods to future climate change scenarios.

Specificity measures the proportion of negatives that are correctly identified (probability of true negatives). An important factor affecting the specificity of the serological and molecular detection methods is the virus population genetic variability and the genetic relationships with other virus populations. Since detection methods are based on specific binding (protein with antibody or nucleic acids with probes or primers), some dissimilar virus variants can fail to be detected giving false negatives. As we discussed in Section 5, altered climatic conditions represent a selection pressure for plant viruses, leading to rapid evolution and therefore to the modification of the genetic composition of plant virus populations. Consequently, increasing genetic divergence of virus populations as compared with standardized antibodies, probes or primers may reduce sensitivity and in extreme cases prevent virus detection (Rubio et al., 2020). At odds with serological and other molecular techniques, HTS enables accurate and unbiased identification of viruses, as it does not require a specific binding that can fail to detect some genetic variants. HTS generates large numbers of nucleotide sequence reads that allow high sensitivity (Santala and Valkonen, 2018). However, bioinformatic processing time and difficulty, and cost, still prevent the routine use of HTS for detection (Maree et al., 2018; Villamor et al., 2019). A potential alternative is a smart design of primers and probes for virus detection, for instance by considering conserved genomic regions during virus evolution as targets for virus detection, which may increase the durability of standard detection methods even under conditions that accelerate virus evolution.

6.2 Chemical and biological control of vectors

As discussed above, plant virus vectors are of concern not only because they are plant pests, but also because of their transmission capacity. Thus, effective strategies need to be implemented to control their populations to minimize plant viral diseases, which usually consist of a combination of chemical, biological, physical and cultural measures (i.e., integrated pest management, IPM), all susceptible to be influenced by climate change. For instance, for chemical or biopesticides, high temperature and low relative humidity cause droplets of the sprayed substances to evaporate faster, which may reduce the amount of product getting in contact with the vector. Temperature may also accelerate the degradation of active compounds (Etheridge et al., 2019). These effects may also lead to sublethal concentrations that may promote the development of resistance in the vector (Andrew and Hill, 2017). Higher temperature and eCO_2 also influence the metabolic activity of the insect that may lead to an increased insecticide detoxification, although this is a double-edge sword and sometimes it may increase the effectiveness of chemical or biocontrol (Amarasekare and Edelson, 2004). IPM often involves biological measures through the use of natural enemies (predators, parasitoids, etc.) (Chandi, 2021). Temperature influences the preferences, abundance, and distribution of parasitoids that suppress aphid vector populations. For instance, survival of the parasitoid *Aphidius colemani*, which parasitizes *M. persicae*, was higher at 20 °C than at 28 °C, determining the abundance of adults (Jerbi-Elayed et al., 2021); and at the higher temperature the parasitoid showed reduced flying capacity, but increased walking capacity (Jerbi-Elayed et al., 2015). Increased temperatures also have been reported to negatively affect parasitoid fecundity (Wang and Keller, 2020). Other examples leading to similar conclusions are summarized in Jones (2016). Despite these negative effects of temperature, it has been shown that parasitoids may quickly adapt to more extreme temperatures (Hance et al., 2007; Jerbi-Elayed et al., 2021), which may allow fast recovery of their capacity to control vector populations. In a triple interaction of aphid-parasitoid-predator, eCO_2 negatively affected the size of the parasitoid population while the opposite effect was observed in the aphid population. However, eCO_2 increased the preference of the parasitoid and the predator for the aphid, resulting in more efficient control of the prey population (Chen et al., 2007). At odds, Wang et al. (2014a) found no significant effects of eCO_2 on *B. tabaci* parasitism by *Encarsia formosa*, with little effect of altered climatic conditions on population dynamics of both partners. These examples illustrate the difficulty in predicting the effects of climate change on the efficiency of natural enemies to control vector populations. More importantly, they highlight that the negative effects on the predator or parasitoid phenology/fitness do not necessarily correlate with predatory or parasitism capacities.

6.3 Genetic resistance and tolerance

A preferred strategy to manage plant virus diseases is the breeding of resistance genes into crop cultivars. Resistance is a quantitative trait defined as the host ability to limit pathogen multiplication (Clarke, 1986; Strauss and Agrawal, 1999). Complete resistance leads to plant immunity, which is the desired trait for plant breeders. Immunity to virus infection is often controlled by single dominant alleles that encode resistance (R) proteins that specifically recognize a sequence or conformational pattern of a virus gene/protein (avirulence determinant, Avr) and induce the programmed death of the infected cells (hypersensitive response, HR), preventing systemic infection (de Ronde et al., 2014; Sacristán and García-Arenal, 2008). Genetic resistance is favored by growers because it may be highly effective and organism-specific and is environmental-friendly (García-Arenal and McDonald, 2003; Rubio et al., 2020). Altered climate may have important negative consequences for the functionality of resistance genes (Jones and Barbetti, 2012). Indeed, warmer temperature may provoke stronger viral infections since R genes are often temperature-sensitive and are inactivated at elevated temperature (Wang et al., 2009). Resistance to *Tobacco mosaic virus* (TMV) conferred by the *N* gene in tomato is a classic example of thermolability. The HR triggered by the recognition of the viral replicase by this gene at 22 °C is inactivated at temperatures above 28 °C (Whitham et al., 1996). Similarly, in wheat resistance to *Wheat streak mosaic virus* conferred by temperature-sensitive *wsm1* and *wsm2* genes was overcame at 32 °C (Tatineni et al., 2016); and resistance to TSWV conferred by the *Tsw* gene becomes inactive at 30 °C (Chung et al., 2018). Alternatively, immunity can be achieved by loss-of-susceptibility conferred by resistance recessive alleles encoding host factors critical for viral infection (Truniger and Aranda, 2009). As for dominant R genes, resistance conferred by recessive alleles is also temperature-sensitive. This is, for instance, the case of the *TOM1* gene that confers resistance to tobamoviruses at 23 °C but not at 36 °C (Yamanaka et al., 2000). Hence, many R genes (dominant and recessive) are likely to become ineffective due to global warming. However, this is not always the case. Examples exist of R genes inducing potent hypersensitive responses at higher temperatures. This is the case of the dominant *Rm* gene that confers resistance to *Potato virus M* in potato cultivars at 28 °C, but not at 20 °C (Tatarowska et al., 2020). Also, a recessive gene of alfalfa conferring resistance to *Alfalfa mosaic virus* has been shown to be temperature-insensitive (Iwai et al., 1992). Research on such thermostable R genes will contribute to adapt disease management strategies based on genetic resistance to future climate change scenarios.

Quantitative resistance (i.e., an incomplete or partial level of resistance that allows certain level of virus multiplication) can be of interest when complete resistance genes are not available. It has been proposed that this partial resistance may be more durable than complete resistance as it imposes a weaker selection pressure on the virus (Pilet-Nayel et al., 2017). Some loci conferring partial resistance have been identified, although it is not known whether temperature modifies their efficacy (Pilet-Nayel et al., 2017). However, the work reviewed in Section 2 on the effects of environmental variables on virus multiplication level suggest that quantitative resistance could be enhanced under climate change, at least in some plant-virus interactions. Unfortunately, the few data on the genetic basis of such quantitative resistance indicates that it is controlled by many genes with minor effects (Caranta et al., 1997; Shukla et al., 2022), and therefore it is of limited agronomical value.

Tolerance is not uncommon in cultivated plants and seems to have evolved in phylogenetically distant species (Pagán and García-Arenal, 2020). Increasing awareness of the relevance of tolerance to pathogens has led to the proposal that tolerance could be a successful strategy to control epidemics in agricultural settings when resistance has not been described, is difficult to breed for, or is quickly overcome by the pathogen. In plant-virus interactions, the term tolerance has been used with different meanings. Cooper and Jones (1983) defined tolerant hosts as those "that a specific virus can infect and in which it can replicate and invade without causing severe symptoms or greatly diminishing the rate or amount of plant growth or marketable yield (or fitness)." Later work pointed to two limitations of this definition: (1) it made it difficult to determine if mild symptoms or no yield/fitness reduction were due to plant mechanisms to cope with the effect of infection (tolerance) or to reduced virus multiplication (resistance), and (2) it defined tolerance as an absolute term, when in nature different degrees of tolerance may occur (Bos and Parlevliet, 1995). These authors proposed that tolerance should be viewed quantitatively as increased yield/fitness in relation to a given virus load, a definition adopted by part of the plant virology community (Alexander et al., 2017; Jeger et al., 2006). These two views of tolerance are not mutually exclusive: the Cooper and Jones (1983) definition would be an extreme case (absolute tolerance) of that of Bos and Parlevliet (1995) in which infection induces no yield/fitness loss at any virus load.

In an agronomical context, absolute tolerance is of higher interest than quantitative (partial) tolerance, mirroring the preference for complete resistance. At odds with complete resistance, absolute tolerance allows virus multiplication at no cost for the host, and it is expected to exert weaker selection pressures than resistance. Hence, it has been proposed to be a more durable control strategy. However, it has some drawbacks that prevented its utilization to date: First, its molecular bases are poorly understood, and when it has been analyzed it seems to have a polygenic control, which is difficult to introgress in cultivated plants (Pagán and García-Arenal, 2018, 2020). Second, as it does not control for virus multiplication, tolerance increases the sources of inoculum, facilitating virus spread to other nontolerant hosts (Montes et al., 2021). As a consequence, tolerance to viruses has been studied only in a few crops, e.g., in tomato, to TYLCV (Lapidot et al., 1997; Rubio et al., 2003), in cereals to BYDV (Foresman et al., 2016; Riedel et al., 2011), in cowpea to *Blackeye cowpea mosaic virus* (Anderson et al., 1996), or in zucchini, to *Zucchini yellow mosaic virus* (Desbiez et al., 2003). However, tolerance would be of interest in crops where virus prevalence is already very high. As higher virus multiplication and transmission rates (which increase prevalence) are predicted consequences of climate change, tolerance might be a promising control strategy in altered climatic conditions.

6.4 Other methods

In addition to R-gene-mediated resistance and to tolerance, plants cope with virus infections through activating their RNA silencing machinery (Ding, 2010), which is induced by dsRNA produced during virus infection. RNA silencing uses 21–24 nucleotides long small interfering RNAs (siRNAs) and microRNAs (miRNAs) to direct sequence-specific cleavage of viral RNAs through the RNA-induced silencing complex (RISC), which can be considered another way to achieve resistance. RNA silencing technology has been successfully applied to target over 60 species of economically important plant viruses, mostly by host genetic modifications (Zhao et al., 2020). However, these transgenic approaches are not only time-consuming and expensive, but also suffer significant regulation and public acceptance issues. To address these limitations and public concerns, exogenous application of naked dsRNA proved to successfully trigger the RNA silencing pathway against viruses (Kaldis et al., 2017; Tenllado and Díaz-Ruíz, 2001; Worrall et al., 2019). However, this strategy does not provide durable protection due to dsRNA instability (up to 20 days after application, incorporating dsRNAs into nanoparticles) (Mitter et al., 2017; Worrall et al., 2019). Despite these technological advances, some reports conclude that RNA silencing or its systemic signaling is inhibited at elevated temperature (Zhong et al., 2013), although studies reporting that RNA silencing is more active at higher temperature and correlates with reduced disease symptoms in infected tissues also exist (Szittya et al., 2003). Hence, its utility in a climate change context remains debatable.

Another method for virus management is cross-protection based on inoculating mild or attenuated viral strains to protect plants against severe strains of the same virus. Cross-protection has been applied to several viruses and crops, such as PepMV in tomato and CTV in citrus crops (Agüero et al., 2018; Pechinger et al., 2019). The mechanism of cross-protection is poorly understood, but all proposed models concur that it seems to be quite species-specific (Jones and Naidu, 2019). To apply cross-protection it is necessary to evaluate the genetic and biological variability of the local virus population and search for mild isolates genetically close to the severe ones (Agüero et al., 2018). Although effective, this strategy poses important risks. First, virus evolution may result in the emergence of highly virulent variants (Fulton, 1986), although some interactions between mild and severe strains remain quite stable over long periods of time (Agüero et al., 2018; Alcaide et al., 2020). Second, the outcome of the infection with the mild isolate may be highly dependent on the environmental conditions, sometimes resulting in synergistic effects (Alcaide et al., 2021). Hence, in a context of climate change, where the rate of virus evolution might be accelerated (see Section 5) and rising temperatures may result in higher multiplication of the milder isolate (Alcaide and Aranda, 2021; Alcaide et al., 2021), cross-protection might not be advisable. Still, for devastating diseases, and when other protection measures fail, benefits of cross-protection may compensate the potential risks.

7. Concluding remarks and future perspectives

From the work summarized here, one can easily derive the conclusion that climate change will have an enormous impact in virtually every aspect of plant-virus interactions.

- To date, most analyses on the effect of climate change on plant virus pathogenicity focused on the effect of a single environmental variable, showing no clear indication whether these conditions are detrimental or beneficial for plant viruses. However, in nature multiple variables can be simultaneously altered. From the few studies addressing the role of various abiotic stresses simultaneously, it seems clear that analyzing them individually does not allow prediction of their combined effects on viral infections. Moreover, in these works a pattern arises: combined higher temperature and eCO_2 generally reduces (or has no effect on) pathogenicity, but the opposite is observed when drought is combined with any of the other abiotic stresses linked to climate change. Hence, understanding the effects of climate change on plant virus pathogenicity requires more analyses considering multiple interactions between environmental variables. In addition, climate change induces multiple alterations in plant physiology, including modifications of C:N ratios, plant hormone defense signaling pathways and RNA silencing machinery. These mechanisms are likely to be simultaneously affected by climate change, but molecular determinants of the environment × plant virus infection interaction have not been addressed from a network perspective. Hence, holistic approaches also must be applied at the molecular level to fully understand the consequences of climate change on virus pathogenicity. Finally, research showing that plants may send "warning" signals when subjected to abiotic stresses shows that the effects of climate change on plant virus pathogenicity might be studied at the levels of both the individual plant and the community.
- Climate change has a large impact on plant virus transmission. For aphids, the most common plant virus vectors, positive and negative effects have been reported depending on the abiotic stress and the plant-virus combination. Apparently, these differences could be explained by the mode of aphid transmission. However, more pathosystems must be explored to support/reject this trend and to understand its molecular basis. For other virus vectors, altered environmental conditions might have contrasting effects, generally positive for whitefly- or thrip-transmitted viruses, and negative for leafhopper- or fungus-transmitted ones. It should also be considered that these effects have been predicted to depend on the standing climate: when climate change has beneficial effects for the vector and the virus transmission in temperate areas, the effect is detrimental in tropical areas, and vice versa. Hence, climate change may modify the relative prevalence of plant viruses depending on their vectors and the geographic

location. More accurate predictions of these effects would require additional work, particularly on the consequences of climate change for viruses transmitted through less common virus vectors (leafhoppers, nematodes, fungi, etc.).

- For at least a quarter of all plant viruses, fitness depends on horizontal and/or of vertical transmission through seeds. In general, the scant work suggests that climate change may favor seed transmission rate, but more studies are needed to confirm this conclusion. Moreover, future studies should consider the effects of abiotic stresses in the number of infected seeds rather than the percentage of infected seeds per plant to have a more comprehensive idea on the link between environmental conditions, seed transmission and virus fitness.
- Climate change may increase contact rates between plant reservoirs and new hosts for virus infection, thereby affecting host jumps and disease emergence. Host extinction events seem to favor plant virus emergence. Although pathogen release has been reported for invasive plant species, there are numerous examples of increased virus prevalence due to invasive species. Note that these two phenomena are not mutually exclusive: invasive plants may be more resistant to infection by native viruses in their naturalized habitats but remain susceptible to viruses they already faced in their natural habitats if these confer benefits to their establishment in new plant communities, which would increase prevalence in the new plant community. Hence, invasion events promoted by climate change are likely to increase contact rates and favor emergence at least for some viruses, with significant negative effects for native plants. A similar rationale can be applied to new interactions between cultivated plants and local flora. Invasions and extinctions also affect the vector populations and those of their natural enemies, but little is known on this aspect. More work on these plants-vector-enemies tritrophic interactions is needed to understand how climate change alters plant virus ecology.
- Even though plant virus mixed infections are common in nature, little is known about the consequences of climate change for these virus-virus interactions. The scarce experimental work suggests that the outcome of the interaction varies depending on the abiotic stress, but the results might be also explained (at least in part) by the nature of the interacting partners (pathogen-pathogen vs pathogen-mutualist). More work on this subject is needed to clarify the relative importance of these factors. Moreover, the interactions between environmental conditions and mixed infections is likely to alter virus transmission by vectors or by seeds,

and these traits also should be considered in future analyses as this may contribute to prediction of epidemics associated with climate change.

- There is little information on how climate change modulates the rate of plant and virus evolution. The scant evidence indicates that altered climate would accelerate mutation rates. Whether this is a general trend and if virus population sizes would increase enough to prevent excessive mutational load is unclear and an interesting avenue for future research. Even less is known on the effects of climate change on the frequency of recombination. Analyses on this aspect are needed to fully understand the potential for virus populations to generate genetic diversity under future climate scenarios. These studies should be combined with analyses on the effect of environmental conditions in modulating the relative importance of genetic drift and selection, which will determine the frequency of virus genotypes generated by mutation and recombination. These studies should consider the nature of the plant-virus interaction in the context of climate change, as depending on the abiotic stress the same plant-virus combination may be pathogenic or mutualistic and might evolve differently. However, to date such analyses are lacking.
- The effect of climate change in almost every aspect of plant-virus interactions is likely to compromise the efficacy of current control measures against virus epidemics. Difficulties for virus detection due to accelerated evolution, inactivation of resistance genes, less effective control based on IPM, or instability of protection conferred by RNA silencing and infection with mild isolates are potential consequences of altered environmental conditions. Hence, research is needed to accommodate existing strategies for disease management to future climatic scenarios. For instance, develop of identification methods based on conserved sequences in viral genomes, identifying temperature-insensitive R genes, or characterizing vector parasitoids and predators adaptable to environmental changes may provide durable control methods. These methods could be complemented with the utilization of novel technologies: remote sensing, artificial intelligence and precision agriculture have great potential, as they will provide tools to adequately manage plants under the environmental challenges they face (including abiotic stresses and pathogens) nearly in real time and at the individual level (Jones and Naidu, 2019). Together with these state-of-the-art technologies, the development of accurate predictive models will allow anticipation of the result of plant-virus-vector-environment interactions, which can be pivotal to face future challenges associated with climate change. Extensive effort has been done in the last decades on this subject

(Jeger, 2020; Jeger et al., 2018; Juroszek and Tiedemann, 2015). Although in this review we did not address in depth the existing mathematical models predicting plant virus disease risk. Accuracy of these models strongly depends on empirical and experimental data used for their parameterization. Filling the knowledge gaps discussed above will contribute to improving their predictive power.

In sum, despite climate change posing a challenge for plant viruses, in many aspects it will provide new opportunities for disease emergence, increasing the need of efficient control strategies. Our understanding of the consequence of climate change for plant virus epidemics has advanced in the last decades at an accelerating pace (Fig. 2), which will allow rapid technological developments to answer this need.

Acknowledgments

This research was funded by Plan Nacional I+D+i, Ministerio de Economía y Competitividad (Agencia Estatal de Investigación), Spain [PID2019-109579RB-I00] to I.P. N.M. was supported by Instituto de Salud Carlos III (RD16/0011/0012).

References

Acosta-Leal, R., Duffy, S., Xiong, Z., Hammond, R.W., Elena, S.F., 2011. Advances in plant virus evolution: translating evolutionary insights into better disease management. Phytopathology 101, 1136–1148.

Agrios, G.N., 2005. Plant Pathology, fifth ed. Elsevier, Amsterdam.

Agüero, J., Gómez-Aix, C., Sempere, R.N., García-Villalba, J., García-Núñez, J., Hernando, Y., Aranda, M.A., 2018. Stable and broad spectrum cross-protection against *Pepino mosaic virus* attained by mixed infection. Front. Plant Sci. 9, 1810.

Aguilar, E., Lozano-Duran, R., 2022. Plant viruses as probes to engineer tolerance to abiotic stress in crops. Stress Biol. https://doi.org/10.1007/s44154-022-00043-4.

Aguilar, E., Allende, L., Del Toro, F.J., Chung, B.N., Canto, T., Tenllado, F., 2015. Effects of elevated CO_2 and temperature on pathogenicity determinants and virulence of potato virus X/Potyvirus-associated synergism. Mol. Plant Microbe Interact. 28, 1364–1373.

Aguilar, E., Del Toro, F.J., Figueira-Galán, D., Hou, W., Canto, T., Tenllado, F., 2020. Virus infection induces resistance to *Pseudomonas syringae* and to drought in both compatible and incompatible bacteria-host interactions, which are compromised under conditions of elevated temperature and CO_2 levels. J. Gen. Virol. 101, 122–135.

Ahmed, W., Simpson, S.L., Bertsch, P.M., Bibby, K., Bivins, A., Blackall, L.L., Bofill-Mas, S., Bosch, A., Brandão, J., Choi, P.M., Ciesielski, M., Donner, E., D'Souza, N., Farnleitner, A.H., Gerrity, D., Gonzalez, R., Griffith, J.F., Gyawali, P., Haas, C.N., Hamilton, K.A., Hapuarachchi, H.C., Harwood, V.J., Haque, R., Jackson, G., Khan, S.J., Khan, W., Kitajima, M., Korajkic, A., La Rosa, G., Layton, B.A., Lipp, E., McLellan, S.L., McMinn, B., Medema, G., Metcalfe, S., Meijer, W.G., Mueller, J.F., Murphy, H., Naughton, C.C., Noble, R.T., Payyappat, S., Petterson, S., Pitkänen, T., Rajal, V.B., Reyneke, B., Roman Jr., F.A., Rose, J.B., Rusiñol, M., Sadowsky, M.J., Sala-Comorera, L., Setoh, Y.X., Sherchan, S.P., Sirikanchana, K., Smith, W., Steele, J.A., Sabburg, R., Symonds, E.M., Thai, P., Thomas, K.V., Tynan, J., Toze, S., Thompson, J., Whiteley, A.S., Wong, J.C.C.,

Sano, D., Wuertz, S., Xagoraraki, I., Zhang, Q., Zimmer-Faust, A.G., Shanks, O.C., 2022. Minimizing errors in RT-PCR detection and quantification of SARS-CoV-2 RNA for wastewater surveillance. Sci. Total Environ. 805, 149877.

Aimone, C.D., Lavington, E., Hoyer, J.S., Deppong, D.O., Mickelson-Young, L., Jacobson, A., Kennedy, G.G., Carbone, I., Hanley-Bowdoin, L., Duffy, S., 2021. Population diversity of cassava mosaic begomoviruses increases over the course of serial vegetative propagation. J. Gen. Virol. 102, 001622.

Albrechtsen, S.E., 2006. Testing Methods for Seed-Transmitted Viruses: Principles and Protocols. CABI, Wallingford.

Alcaide, C., Aranda, M.A., 2021. Determinants of persistent patterns of *Pepino mosaic virus* mixed infections. Front. Microbiol. 12, 694492.

Alcaide, C., Rabadán, M.P., Juárez, M., Gómez, P., 2020. Long-Term cocirculation of two strains of *Pepino mosaic virus* in tomato crops and its effect on population genetic variability. Phytopathology 110, 49–57.

Alcaide, C., Sardanyés, J., Elena, S.F., Gómez, P., 2021. Increasing temperature alters the within-host competition of viral strains and influences virus genetic variability. Virus Evol. 7, veab017.

Alexander, H.M., Mauck, K.E., Whitfield, A.E., Garrett, K.A., Malmstrom, C.M., 2014. Plant-virus interactions and the agro-ecological interface. Eur. J. Plant Pathol. 138, 529–547.

Alexander, H.M., Bruns, E., Schebor, H., Malmstrom, C.M., 2017. Crop-associated virus infection in a native perennial grass: reduction in plant fitness and dynamic patterns of virus detection. J. Ecol. 105, 1021–1031.

Ali, S., Khan, M.A., Habib, A., Rasheed, S., Iftikhar, Y., 2005. Correlation of environmental conditions with *Okra yellow vein mosaic virus* and *Bemisia tabaci* population density. Int. J. Agric. Biol. 7, 142–144.

Ali, A., Li, H., Schneider, W.L., Sherman, D.J., Gray, S., Smith, D., Roossinck, M.J., 2006. Analysis of genetic bottlenecks during horizontal transmission of *Cucumber mosaic virus*. J. Virol. 80, 8345–8350.

Amarasekare, K.G., Edelson, J.V., 2004. Effect of temperature on efficacy of insecticides to differential grasshopper (Orthoptera: Acrididae). J. Econ. Entomol. 97, 1595–1602.

Amari, K., Huang, C., Heinlein, M., 2021. Potential impact of global warming on virus propagation in infected plants and agricultural productivity. Front. Plant Sci. 12, 649768.

Anderson, R.M., May, R., 1982. Coevolution of hosts and parasites. Parasitology 85, 411–426.

Anderson, E.J., Kline, A.S., Morelock, T.E., McNew, R.W., 1996. Tolerance to *Blackeye cowpea mosaic potyvirus* not correlated with decreased virus accumulation or protection from cowpea stunt disease. Plant Dis. 80, 847–852.

Anderson, P.K., Cunningham, A.A., Patel, N.G., Morales, F.J., Epstein, P.R., Daszak, P., 2004. Emerging infectious diseases of plants: pathogen pollution, climate change and agrotechnology drivers. Trends Ecol. Evol. 19, 535–544.

Andrew, N.R., Hill, S.J., 2017. Effect of climate change on insect pest management. In: Coll, M., Wajnberg, E. (Eds.), Environmental Pest Management: Challenges for Agronomists, Ecologists, Economists and Policymakers. John Wiley & Sons Ltd., Hoboken, pp. 197–223.

Anwar, N., Ahmad, I., Raja, M.A.Z., Naz, S., Shoaib, M., Kiani, A.K., 2022. Artificial intelligence knacks-based stochastic paradigm to study the dynamics of plant virus propagation model with impact of seasonality and delays. Eur. Phys. J. Plus 137, 144.

Aregbesola, O.Z., Legg, J.P., Sigsgaard, L., Lund, O.S., Rapisarda, C., 2019. Potential impact of climate change on whiteflies and implications for the spread of vectored viruses. J. Pest Sci. 92, 381–392.

Barreto, S.S., Hallwass, M., Aquino, O.M., Inoue-Nagata, A.K., 2013. A study of weeds as potential inoculum sources for a tomato-infecting begomovirus in central Brazil. Phytopathology 103, 436–444.
Belfield, E.J., Brown, C., Ding, Z.J., Chapman, L., Luo, M., Hinde, E., van Es, S.W., Johnson, S., Ning, Y., Zheng, S.J., Mithani, A., Harberd, N.P., 2021. Thermal stress accelerates *Arabidopsis thaliana* mutation rate. Genome Res. 31, 40–50.
Bergès, S.E., Vile, D., Vazquez-Rovere, C., Blanc, S., Yvon, M., Bediee, A., Rolland, G., Dauzat, M., van Munster, M., 2018. Interactions between drought and plant genotype change epidemiological traits of *Cauliflower mosaic virus*. Front. Plant Sci. 9, 703.
Bergès, S.E., Vile, D., Yvon, M., Masclef, D., Dauzat, M., van Munster, M., 2021. Water deficit changes the relationships between epidemiological traits of *Cauliflower mosaic virus* across diverse *Arabidopsis thaliana* accessions. Sci. Rep. 11, 24103.
Bertschinger, L., Bühler, L., Dupuis, B., Duffy, B., Gessler, C., Forbes, G.A., Keller, E.R., Scheidegger, U.C., Struik, P.C., 2017. Incomplete infection of secondarily infected potato plants—an environment dependent underestimated mechanism in plant virology. Front. Plant Sci. 8, 74.
Betancourt, M., Fereres, A., Fraile, A., García-Arenal, F., 2008. Estimation of the effective number of founders that initiate an infection after aphid transmission of a multipartite plant virus. J. Virol. 82, 12416–12421.
Boccardo, G., Lisa, V., Luisoni, E., Milne, R.G., 1987. Cryptic plant viruses. Adv. Virus Res. 32, 171–214.
Bos, L., Parlevliet, J.E., 1995. Concepts and terminology on plant-pest relationships: toward consensus in plant pathology and crop protection. Annu. Rev. Phytopathol. 33, 69–102.
Bosque-Perez, N.A., Eigenbrode, S.D., 2011. The influence of virus-induced changes in plants on aphid vectors: insights from luteovirus pathosystems. Virus Res. 159, 201–205.
Bosso, L., Di Febbraro, M., Cristinzio, G., Zoina, A., Russo, D., 2016. Shedding light on the effects of climate change on the potential distribution of *Xylella fastidiosa* in the Mediterranean basin. Biol. Invasions 18, 1759–1768.
Bradamante, G., Mittelsten Scheid, O., Incarbone, M., 2021. Under siege: virus control in plant meristems and progeny. Plant Cell 33, 2523–2537.
Braendle, C., Davis, G.K., Brisson, J.A., Stern, D.L., 2006. Wing dimorphism in aphids. Heredity 97, 192–199.
Brault, V., Uzest, M., Monsion, B., Jacquot, E., Blanc, S., 2010. Aphids as transport devices for plant viruses. C. R. Biol. 333, 524–538.
Brisson, J.A., 2010. Aphid wing dimorphisms: linking environmental and genetic control of trait variation. Philos. Trans. R. Soc. Lond. B Biol. Sci. 365, 605–616.
Brunner, M.I., Swain, D.L., Wood, R.R., Willkofer, F., Done, J.M., Gilleland, E., Ludwig, R., 2021. An extremeness threshold determines the regional response of floods to changes in rainfall extremes. Commun. Earth Environ. 2, 173.
Bueso, E., Serrano, R., Pallás, V., Sánchez-Navarro, J.A., 2017. Seed tolerance to deterioration in arabidopsis is affected by virus infection. Plant Physiol. Biochem. 116, 1–8.
Bujarski, J., 2013. Genetic recombination in plant-infecting messenger-sense RNA viruses: overview and research perspectives. Front. Plant Sci. 4, 68.
Calleja-Cabrera, J., Boter, M., Oñate-Sánchez, L., Pernas, M., 2020. Root growth adaptation to climate change in crops. Front. Plant Sci. 11, 544.
Campbell, R.N., 1996. Fungal transmission of plant viruses. Annu. Rev. Phytopathol. 34, 87–108.
Canto, T., Palukaitis, P., 2002. Novel N gene-associated, temperature-independent resistance to the movement of *Tobacco mosaic virus* vectors neutralized by a *Cucumber mosaic virus* RNA1 transgene. J. Virol. 76, 12908–12916.
Canto, T., Aranda, M.A., Fereres, A., 2009. Climate change effects on physiology and population processes of hosts and vectors that influence the spread of hemipteran-borne plant viruses. Glob. Chang. Biol. 15, 1884–1894.

Caranta, C., Lefebvre, V., Palloix, A., 1997. Polygenic resistance of pepper to potyviruses consists of a combination of isolate-specific and broad-spectrum quantitative trait loci. Mol. Plant Microbe Interact. 10, 872–878.
Carmo-Sousa, M., Moreno, A., Garzo, E., Fereres, A., 2014. A non-persistently transmitted-virus induces a pull-push strategy in its aphid vector to optimize transmission and spread. Virus Res. 186, 38–46.
Carrasco, P., de la Iglesia, F., Elena, S.F., 2007. Distribution of fitness and virulence effects caused by single-nucleotide substitutions in *Tobacco etch virus*. J. Virol. 81, 12979–12984.
Castillo, P., Gomez-Barcina, A., Jimenez-Diaz, R.M., 1996. Plant parasitic nematodes associated with chickpea in Southern Spain and effect of soil temperature on reproduction of *Pratylenchus thornei*. Nematologica 42, 211–219.
Cesarz, S., Reich, P.B., Scheu, S., Ruess, L., Schaefer, M., Eisenhauer, N., 2015. Nematode functional guilds, not trophic groups, reflect shifts in soil food webs and processes in response to interacting global change factors. Pedologia 58, 23–32.
Chakravarty, A., Reddy, V.S., Rao, A.L.N., 2020. Unravelling the stability and capsid dynamics of the three virions of *Brome mosaic virus* assembled autonomously *in vivo*. J. Virol. 94, e01794-19.
Chandi, R.S., 2021. Integrated management of insect vectors of plant pathogens. Agric. Rev. 42, 87–92.
Chao, L., 1990. Fitness of RNA virus decreased by Muller's ratchet. Nature 348, 454–455.
Chappell, T.M., Beaudoin, A.L.P., Kennedy, G.G., 2013. Interacting virus abundance and transmission intensity underlie *Tomato spotted wilt virus* incidence: an example weather-based model for cultivated tobacco. PLoS One 8, e73321.
Chellappan, P., Vanitharani, R., Ogbe, F., Fauquet, C.M., 2005. Effect of temperature on geminivirus-induced RNA silencing in plants. Plant Physiol. 138, 1828–1841.
Chen, F.J., MacFarlane, S.A., Wilson, T.M.A., 1995. Effect of cultivation temperature on the spontaneous development of deletions in soilborne *Wheat mosaic furovirus* RNA 2. Phytopathology 85, 299–306.
Chen, F.J., Wu, G., Parajulee, M.N., Feng, G., 2007. Impact of elevated CO_2 on the third trophic level: a predator *Harmonia axyridis* and a parasitoid *Aphidius picipes*. Biocontrol Sci. Technol. 17, 313–324.
Choi, K.S., Toro, F., Tenllado, F., Canto, T., Chung, B.N., 2017. A model to explain temperature dependent systemic infection of potato plants by *Potato virus Y*. Plant Pathol. J. 33, 206–211.
Chung, B.N., Choi, K.S., Ahn, J.J., Joa, J.H., Do, K.S., Park, K.S., 2015. Effects of temperature on systemic infection and symptom expression of *Turnip mosaic virus* in Chinese cabbage (*Brassica campestris*). Plant Pathol. J. 31, 363–370.
Chung, B.N., Koh, S.W., Choi, K.S., Joa, J.H., Kim, C.H., Selvakumar, G., 2017. Temperature and CO_2 level influence *Potato leafroll virus* infection in *Solanum tuberosum*. Plant Pathol. J. 33, 522–527.
Chung, B., Lee, J.H., Kang, B.C., Koh, S.W., Joa, J.H., Choi, K.S., Ahn, J.J., 2018. HR-mediated defense response is overcome at high temperatures in capsicum species. Plant Pathol. J. 34, 71–77.
Clarke, D.D., 1986. Tolerance of parasites and disease in plants and its significance in host-parasite interactions. Adv. Plant Pathol. 5, 161–198.
Claudel, P., Chesnais, Q., Fouche, Q., Krieger, C., Halter, D., Bogaert, F., Meyer, S., Boissinot, S., Hugueney, P., Ziegler-Graff, V., Ameline, A., Brault, V., 2018. The aphid-transmitted *Turnip yellows virus* differentially affects volatiles emission and subsequent vector behavior in two *Brassicaceae* plants. Int. J. Mol. Sci. 19, 2316.
Cobos, A., Montes, N., López-Herranz, M., Gil-Valle, M., Pagán, I., 2019. Within-host multiplication and speed of colonization as infection traits associated with plant virus vertical transmission. J. Virol. 93, e01078-19.

Collins, M., Knutti, R., Arblaster, J., Dufresne, J.-L., Fichefet, T., Friedlingstein, P., Gao, X., Gutowski, W.J., Johns, T., Krinner, G., Shongwe, M., Tebaldi, C., Weaver, A.J., Wehner, M., 2013. Long-term climate change: projections, commitments and irreversibility. In: Stocker, T.F., Qin, D., Plattner, G.-K., Tignor, M., Allen, S.K., Boschung, J., Midgley, P.M. (Eds.), Climate Change 2013: The Physical Science Basis. Contribution of Working Group I to the Fifth Assessment Report of the Intergovernmental Panel on Climate Change. Cambridge University Press, Cambridge, pp. 1029–1133.

Cooper, J.I., Jones, A.T., 1983. Responses of plants to viruses: proposals for the use of terms. Phytopathology 73, 127–128.

Coutts, B.A., Prince, R.T., Jones, R.A., 2009. Quantifying effects of seedborne inoculum on virus spread, yield losses, and seed infection in the *Pea seed-borne mosaic virus*-field pea pathosystem. Phytopathology 99, 1156–1167.

Culbreath, A.K., Srinivasan, R., 2011. Epidemiology of spotted wilt disease of peanut caused by *Tomato spotted wilt virus* in the southeastern U.S. Virus Res. 159, 101–109.

Cumutte, L.V., Simmons, A.M., Abd-Rabbou, S., 2014. Climate change and *Bemisia tabaci* (Hemiptera: Aleyrodidae): impacts of temperature and carbon dioxide on life history. Ann. Entomol. Soc. Am. 107, 933–943.

Dáder, B., Fereres, A., Moreno, A., Trebicki, P., 2016. Elevated CO_2 impacts bell pepper growth with consequences to *Myzus persicae* life history, feeding behavior and virus transmission ability. Sci. Rep. 6, 19120.

Dai, Z.-C., Zhu, B., Wan, J.S.H., Rutherford, S., 2022. Editorial: global changes and plant invasions. Front. Ecol. Evol. 10, 845816.

de Ronde, D., Butterbach, P., Kormelink, R., 2014. Dominant resistance against plant viruses. Front. Plant Sci. 5, 307.

Debrot, E.A., 1964. Studies on a strain of raspberry ringspot virus occurring in England. Ann. Appl. Biol 54, 183–191.

Del Toro, F.J., Aguilar, E., Hernández-Walias, F.J., Tenllado, F., Chung, B.-N., Canto, T., 2015. High temperature, high ambient CO_2 affect the interactions between three positive-sense RNA viruses and a compatible host differentially, but not their silencing suppression efficiencies. PLoS One 10, e0136062.

Del Toro, F.J., Rakhshandehroo, F., Larruy, B., Aguilar, E., Tenllado, F., Canto, T., 2017. Effects of simultaneously elevated temperature and CO_2 levels on *Nicotiana benthamiana* and its infection by different positive-sense RNA viruses are cumulative and virus type-specific. Virology 511, 184–192.

Desbiez, C., Gal-On, A., Girard, M., WipfScheibel, C., Lecoq, H., 2003. Increase in *Zucchini yellow mosaic virus* symptom severity in tolerant zucchini cultivars is related to a point mutation in P3 protein and is associated with a loss of relative fitness on susceptible plants. Phytopathology 93, 1478–1484.

Diaz, B.M., Fereres, A., 2005. Life table and population parameters of *Nasonovia ribisnigri* (Homoptera: Aphididae) at different constant temperatures. Environ. Entomol. 34, 527–534.

Dietzgen, R.G., Mann, K.S., Johnson, K.N., 2016. Plant virus-insect vector interactions: current and potential future research directions. Viruses 8, 303.

Diffenbaugh, N.S., Swain, D.L., Touma, D., 2015. Anthropogenic warming has increased drought risk in California. Proc. Natl. Acad. Sci. U. S. A. 112, 3931–3936.

Ding, S.-W., 2010. RNA-based antiviral immunity. Nat. Rev. Immunol. 10, 632–644.

Drake, J.W., Charlesworth, B., Charlesworth, D., Crow, J.F., 1998. Rates of spontaneous mutation. Genetics 148, 1667–1686.

Dreissig, S., Mascher, M., Heckmann, S., 2019. Variation in recombination rate is shaped by domestication and environmental conditions in barley. Mol. Biol. Evol. 36, 2029–2039.

Duffy, S., Shackelton, L.A., Holmes, E.C., 2008. Rates of evolutionary change in viruses: patterns and determinants. Nat. Rev. Genet. 9, 267–276.

Dwyer, G.I., Gibbs, M.J., Gibbs, A.J., Jones, R.A.C., 2007. *Wheat streak mosaic* virus in Australia: relationship to isolates from the Pacific Northwest of the USA and its dispersion via seed transmission. Plant Dis. 91, 164–170.

Ebi, K.L., Vanos, J., Baldwin, J.W., Bell, J.E., Hondula, D.M., Errett, N.A., Hayes, K., Reid, C.E., Saha, S., Spector, J., Berry, P., 2021. Extreme weather and climate change: population health and health system implications. Annu. Rev. Public Health 42, 293–315.

Eigenbrode, S.D., Ding, H., Shiel, P., Berger, P.H., 2002. Volatiles from potato plants infected with *Potato leafroll virus* attract and arrest the virus vector, *Myzus persicae* (Homoptera, Aphididae). Proc. Biol. Sci. 269, 455–460.

Elena, S.F., Sanjuán, R., 2005. On the adaptive value of high mutation rates in RNA viruses: separating causes from consequences. J. Virol. 79, 11555–11558.

Elena, S.F., Bedhomme, S., Carrasco, P., Cuevas, J.M., de la Iglesia, F., Lafforgue, G., Lalić, J., Pròsper, A., Tromas, N., Zwart, M.P., 2011. The evolutionary genetics of emerging plant RNA viruses. Mol. Plant Microbe Interact. 24, 287–293.

Elena, S.F., Fraile, A., García-Arenal, F., 2014. Evolution and emergence of plant viruses. Adv. Virus Res. 88, 161–191.

Etheridge, B., Gore, J., Catchot, A.L., Cook, D.R., Musser, F.R., Larson, E.J., 2019. Influence of temperature on the efficacy of foliar insecticide sprays against sugarcane aphid (Hemiptera: aphididae) populations in grain sorghum. J. Econ. Entomol. 112, 196–200.

Fabre, F., Plantegenest, M., Mieuzet, L., Dedryver, C.A., Leterrier, J.-L., Jacquot, E., 2005. Effects of climate and land use on the occurrence of viruliferous aphids and the epidemiology of barley yellow dwarf disease. Agric. Ecosyst. Environ. 106, 49–55.

Fabre, F., Moury, B., Johansen, E.I., Simon, V., Jacquemond, M., 2014. Narrow bottlenecks affect *Pea seedborne mosaic virus* populations during vertical seed transmission but not during leaf colonization. PLoS Pathog. 10, e1003833.

Fatnassi, H., Pizzol, J., Senoussi, R., Biondi, A., Desneux, N., Poncel, C., Boular, T., 2015. Within-crop air temperature and humidity outcomes on spatio-temporal distribution of the key rose pest *Frankliniella occidentalis*. PLoS One 10, e0126655.

Fawcett, H.S., 1931. The importance of investigations of the effects of known mixtures of microorganisms. Phytopathology 21, 545–550.

Fereres, A., Moreno, A., 2009. Behavioural aspects influencing plant virus transmission by homopteran insects. Virus Res. 141, 158–168.

Fereres, A., Raccah, B., 2015. Plant Virus Transmission by Insects. John Wiley & Sons Ltd., Hoboken.

Fereres, A., Kampmeier, G.E., Irwin, M.E., 1999. Aphid attraction and preference for soybean and pepper plants infected with *Potyviridae*. Ann. Entomol. Soc. Am. 92, 542–548.

Fones, H.N., Gurr, S.J., 2017. NO_Xious gases and the unpredictability of emerging plant pathogens under climate change. BMC Biol. 15, 36.

Foresman, B.J., Oliver, R.E., Jackson, E.W., Chao, S., Arruda, M.P., Kolb, F.L., 2016. Genome-wide association mapping of *Barley yellow dwarf virus* tolerance in spring oat (*Avena sativa* L.). PLoS One 11, e0155376.

Fraile, A., Hily, J.M., Pagán, I., Pacios, L.F., García-Arenal, F., 2014. Host resistance selects for traits unrelated to resistance-breaking that affect fitness in a plant virus. Mol. Biol. Evol. 31, 928–939.

Fulton, R.W., 1986. Practices and precautions in the use of cross protection for plant virus disease control. Annu. Rev. Phytopathol. 24, 67–81.

Futuyama, D.J., Moreno, G., 1988. The evolution of ecological specialization. Annu. Rev. Ecol. Syst. 19, 207–233.

Gadhave, K.R., Gautam, S., Rasmussen, D.A., Srinivasan, R., 2020. Aphid transmission of potyvirus: the largest plant-infecting RNA virus genus. Viruses 12, 773.

Gamarra, H., Carhuapoma, P., Cumapa, L., González, G., Muñoz, J., Sporleder, M., Kreuze, J., 2020a. A temperature-driven model for *Potato yellow vein virus* transmission efficacy by *Trialeurodes vaporariorum* (Hemiptera: Aleyrodidae). Virus Res. 289, 198109.

Gamarra, H., Sporleder, M., Carhuapoma, P., Kroschel, J., Kreuze, J., 2020b. A temperature-dependent phenology model for the greenhouse whitefly *Trialeurodes vaporariorum* (Hemiptera: Aleyrodidae). Virus Res. 289, 198107.

García-Arenal, F., Fraile, A., 2013. Trade-offs in host range evolution of plant viruses. Plant Pathol. 62, 2–9.

García-Arenal, F., McDonald, B.A., 2003. An analysis of the durability of resistance to plant viruses. Phytopathology 93, 941–952.

García-Arenal, F., Zerbini, F.M., 2019. Life on the edge: geminiviruses at the interface between crops and wild plant hosts. Annu. Rev. Virol. 6, 411–433.

García-Arenal, F., Fraile, A., Malpica, J.M., 2001. Variability and genetic structure of plant virus populations. Annu. Rev. Phytopathol. 39, 157–186.

García-Cano, E., Resende, R.O., Fernández-Muñoz, R., Moriones, E., 2006. Synergistic interaction between *Tomato chlorosis virus* and *Tomato spotted wilt virus* results in breakdown of resistance in tomato. Phytopathology 96, 1263–1269.

Garrett, K.A., Dendy, S.P., Frank, E.E., Rouse, M.N., Travers, S.E., 2006. Climate change effects on plant disease: genomes to ecosystems. Annu. Rev. Phytopathol. 44, 489–509.

Gerrard, M.G., 2013. Effects of WClMV and Root Flooding on White Clover. MSc Thesis, University of Otago, New Zealand.

Gharbi, S., Verhoyen, M., 1994. Effect of temperature and pH on *Olpidium brassicae*. In: Mededelingen—Faculteit Landbouwkundige en Toegepaste Biologische Wetenschappen, Universiteit Gent, vol. 59, pp. 819–833.

Gilbertson, R.L., Batuman, O., Webster, C.G., Adkins, S., 2015. Role of insect supervectors *Bemisia tabaci* and *Frankliniella occidentalis* in the emergence and global spread of plant viruses. Annu. Rev. Phytopathol. 2, 67–93.

Gómez, P., Sempere, R.N., Elena, S.F., Aranda, M.A., 2009. Mixed infections of *Pepino mosaic virus* strains modulate the evolutionary dynamics of this emergent virus. J. Virol. 83, 12378–12387.

González, R., Butković, A., Elena, S.F., 2020. From foes to friends: viral infections expand the limits of host phenotypic plasticity. Adv. Virus Res. 106, 85–121.

González, R., Butković, A., Escaray, F.J., Martínez-Latorre, J., Melero, Í., Pérez-Parets, E., Gómez-Cadenas, A., Carrasco, P., Elena, S.F., 2021. Plant virus evolution under strong drought conditions results in a transition from parasitism to mutualism. Proc. Natl. Acad. Sci. U. S. A. 118, e2020990118.

Grimm, N.B., Chapin, F.S., Bierwagen, B., Gonzalez, P., Groffman, P.M., Luo, Y., Melton, F., Nadelhoffer, K., Pairis, A., Raymond, P.A., Schimel, J., Williamson, C.E., 2013. The impacts of climate change on ecosystem structure and function. Front. Ecol. Environ. 11, 474–482.

Grogan, R.G., 1980. Control of lettuce mosaic with virus. Plant Dis. 64, 446.

Guerret, M.G.L., Barbetti, M.J., You, M.P., Jones, R.C.A., 2016. Effects of temperature on disease severity in plants of subterranean clover infected singly or in mixed infection with *Bean yellow mosaic virus* and *Kabatiella caulivora*. J. Phytopathol. 164, 608–619.

Guo, H., Huang, L., Sun, Y., Guo, H., Ge, F., 2016. The contrasting effects of elevated co_2 on TYLCV infection of tomato genotypes with and without the resistance gene, *mi*-1.2. Front. Plant Sci. 7, 1680.

Guo, H., Wan, S., Ge, F., 2017. Effect of elevated CO_2 and O_3 on phytohormone-mediated plant resistance to vector insects and insect-borne plant viruses. Sci. China Life Sci. 60, 816–825.

Guo, H., Ge, P., Tong, J., Zhang, Y., Peng, X., Zhao, Z., Ge, F., Sun, Y., 2021. Elevated carbon dioxide levels decreases *Cucumber mosaic virus* accumulation in correlation with greater accumulation of rgs-CaM, an inhibitor of a viral suppressor of RNAi. Plants 10, 59.

Gutiérrez, S., Michalakis, Y., Blanc, S., 2012. Virus population bottlenecks during within-host progression and host-to-host transmission. Curr. Opin. Virol. 2, 546–555.

Hamada, T., Mise, K., Kiba, A., Hikichi, Y., 2019. Systemic necrosis in tomato induced by a Japanese isolate of *Rehmannia mosaic virus* in a temperature-sensitive manner. Plant Pathol. 68, 1025–1032.

Hamelin, F.M., Allen, L.J.S., Prendeville, H.R., Hajimorad, M.R., Jeger, M.J., 2016. The evolution of plant virus transmission pathways. J. Theor. Biol. 396, 75–89.

Hance, T., van Baaren, J., Vernon, P., Boivin, G., 2007. Impact of extreme temperatures on parasitoids in a climate change perspective. Annu. Rev. Entomol. 52, 107–126.

Harrison, B.D., 1956. A strain of *Tobacco mosaic virus* infecting plantago spp. in Scotland. Phytopathology 5, 147–148.

Hataya, T., 2021. An improved method for the extraction of nucleic acids from plant tissue without grinding to detect plant viruses and viroids. Plants 10, 2683.

Heagle, A.S., 2003. Influence of elevated carbon dioxide on interactions between *Frankliniella occidentalis* and *Trifolium repens*. Environ. Entomol. 32, 421–424.

Hillung, J., Cuevas, J.M., Elena, S.F., 2015. Evaluating the within-host fitness effects of mutations fixed during virus adaptation to different ecotypes of a new host. Philos. Trans. R. Soc. Lond. B Biol. Sci. 370, 20140292.

Hobbs, H.A., McLaughlin, M.R., 1990. A non-aphid-transmissible isolate of *Bean yellow mosaic virus*-Scott that is transmissible from mixed infections with *Pea mosaic virus*-204-1. Phytopathology 80, 268–272.

Hogenhout, S.A., Ammar el, D., Whitfield, A.E., Redinbaugh, M.G., 2008. Insect vector interactions with persistently transmitted viruses. Annu. Rev. Phytopathol. 46, 327–359.

von Holle, B., Weber, S.E., Nickerson, D.M., 2020. The influence of warming and biotic interactions on the potential for range expansion of native and nonnative species. AoB Plants 12, plaa040.

Holmes, E.C., 2009. The Evolution and Emergence of RNA Viruses. Oxford University Press, Oxford.

Hudson, P., Perkins, S., Cattadori, I., 2008. The emergence of wildlife disease and the application of ecology. In: Ostfeld, R.S., Keesing, F., Eviner, V.T. (Eds.), Infectious Disease Ecology: Effects of Ecosystems on Disease and of Disease on Ecosystems. Princeton University Press, Princeton, pp. 347–367.

Hull, R., 2014. Plant Virology, fifth ed. Academic Press, London.

IPCC, 2014. Climate Change 2014: Synthesis Report. Contribution of Working Groups I, II and III to the Fifth Assessment Report of the Intergovernmental Panel on Climate Change. IPCC, Geneva.

Islam, W., Noman, A., Naveed, H., Alamri, S.A., Hashem, M., Huang, Z., Chen, H.Y.H., 2020. Plant-insect vector-virus interactions under environmental change. Sci. Total Environ. 701, 135044.

Iwai, H., Horikawa, Y., Figueiredo, G., Hiruki, C., 1992. A recessive resistance gene of alfalfa to *Alfalfa mosaic virus* and impact of high ambient temperature on virus resistance. Plant Pathol. 41, 69–75.

Jamieson, M.A., Burkle, L.A., Manson, J.S., Runyon, J.B., Trowbridge, A.M., Zientek, J., 2017. Global change effects on plant-insect interactions: the role of phytochemistry. Curr. Opin. Insect Sci. 23, 70–80.

Jež-Krebelj, A., Rupnik- Cigoj, M., Stele, M., Chersicola, M., Pompe-Novak, M., Sivilotti, P., 2022. The physiological impact of GFLV virus infection on grapevine water status: first observations. Plants 11, 161.
Jeger, M.J., 2020. The epidemiology of plant virus disease: towards a new synthesis. Plants 9, 1768.
Jeger, M.J., Seal, S.E., van den Bosch, F., 2006. Evolutionary epidemiology of plant virus disease. Adv. Virus Res. 67, 163–203.
Jeger, M.J., Madden, L.V., van den Bosch, F., 2018. Plant virus epidemiology: applications and prospects for mathematical modeling and analysis to improve understanding and disease control. Plant Dis. 102, 837–854.
Jennings, M.D., Harris, G.M., 2017. Climate change and ecosystem composition across large landscapes. Landsc. Ecol. 2017 (32), 195–207.
Jerbi-Elayed, M., Grissa-Lebdi, K., Le Goff, G., Hance, T., 2015. Influence of temperature on flight, walking and oviposition capacities of two aphid parasitoid species (hymenoptera: aphidiinae). J. Insect Behav. 28, 157–166.
Jerbi-Elayed, M., Foray, V., Tougeron, K., Grissa-Lebdi, K., Hance, T., 2021. Developmental temperature affects life-history traits and heat tolerance in the aphid parasitoid *Aphidius colemani*. Insects 12, 852.
Jiang, S., Zhang, N., Wang, S., Li, J., Zhang, B., Zheng, C., 2014. Effects of heat shock and life parameters of *Frankliniella occidentalis* (Thysanoptera: Thripidae) F1 offspring. Fla. Entomol. 97, 1157–1166.
Johansen, E., Edwards, M.C., Hampton, R.O., 1994. Seed transmission of viruses: current perspectives. Annu. Rev. Phytopathol. 32, 363–386.
Johnson, J., 1921. The relation of air temperature to certain plant diseases. Phytopathology 11, 446–458.
Jones, R.A.C., 2009. Plant virus emergence and evolution: origins, new encounter scenarios, factors driving emergence, effects of changing world conditions, and prospects for control. Virus Res. 141, 113–130.
Jones, R.A.C., 2016. Future scenarios for plant virus pathogens as climate change progresses. Adv. Virus Res. 95, 87–147.
Jones, R.A.C., 2020. Disease pandemics and major epidemics arising from new encounters between indigenous viruses and introduced crops. Viruses 12, 1388.
Jones, R.A.C., Barbetti, M.J., 2012. Influence of climate change on plant disease infections and epidemics caused by viruses and bacteria. CAB Rev. 7, 1–32.
Jones, R.C.A., Naidu, R.A., 2019. Global dimensions of plant virus diseases: current status and future perspectives. Annu. Rev. Virol. 6, 20.1–20.23.
Jones, R.A.C., Proudlove, W., 1991. Further studies on cucumber mosaic virus infection of narrow-leafed lupin (*Lupinus angustifolius*): seed-borne infection, aphid transmission, spread and effects on grain yield. Ann. Appl. Biol. 118, 319–329.
Joseph, M.B., Mihaljevic, J.R., Orlofske, S.A., Paull, S.H., 2013. Does life history mediate changing disease risk when communities disassemble? Ecol. Lett. 16, 1405–1412.
Juroszek, P., Tiedemann, A.v., 2015. Linking plant disease models to climate change scenarios to project future risks of crop diseases: a review. J. Plant Dis. Protect. 122, 3–15.
Juroszek, P., Racca, P., Link, S., Farhumand, J., Kleinhenz, B., 2020. Overview on the review articles published during the past 30 years relating to the potential climate change effects on plant pathogens and crop disease risks. Plant Pathol. 69, 179–193.
Kaldis, A., Berbati, M., Melita, O., Reppa, C., Holeva, M., Otten, P., Voloudakis, A., 2017. Exogenously applied dsRNA molecules deriving from the *Zucchini yellow mosaic virus* (ZYMV) genome move systemically and protect cucurbits against ZYMV. Mol. Plant Pathol. 19, 883–895.

Kanakala, S., Kontsedalov, S., Lebedev, G., Ghanim, M., 2019. Plant-mediated silencing of the whitefly *Bemisia tabaci* cyclophilin B and heat shock protein 70 impairs insect development and virus transmission. Front. Physiol. 10, 557.

Karpicka-Ignatowska, K., Laska, A., Rector, B.G., Skoracka, A., Kuczyński, L., 2021. Temperature-dependent development and survival of an invasive genotype of wheat curl mite, *Aceria tosichella*. Exp. Appl. Acarol. 83, 513–525.

Kassanis, B., 1957. Effects of changing temperature on plant virus diseases. Adv. Virus Res. 4, 221–241.

Keesing, F., Belden, L.K., Daszk, P., Dobson, A., Harwell, C.D., Holt, R.D., Hudson, P., Jolles, A., Jones, K.E., Mitchell, C.E., Myers, S.S., Bogich, T., Ostfeld, R.S., 2010. Impacts of biodiversity on the emergence and transmission of infectious diseases. Nature 468, 647–652.

King, A.M., Lefkowitz, E., Adams, M.J., Carstens, E.B., 2012. Virus Taxonomy: Ninth Report of the International Committee on Taxonomy of Viruses. Elsevier, Amsterdam.

Kirchner, S.M., Hiltunen, L., Döring, T.F., Virtanen, E., Palohuhta, J.P., Valkonen, J.P., 2013. Seasonal phenology and species composition of the aphid fauna in a northern crop production area. PLoS One 8, e71030.

Kotakis, C., Vrettos, N., Daskalaki, M.G., Kotzabasis, K., Kalantidis, K., 2011. DCL3 and DCL4 are likely involved in the light intensity-RNA silencing cross talk in *Nicotiana benthamiana*. Plant Signal. Behav. 6, 1180–1182.

Kriticos, D.J., Darnell, R.E., Yonow, T., Ota, N., Sutherst, R.W., Parry, H.R., Mugerwa, H., Maruthi, M.N., Seal, S.E., Colvin, J., Macfadyen, S., Kalyebi, A., Hulthen, A., De Barro, P.J., 2020. Improving climate suitability for *Bemisia tabaci* in East Africa is correlated with increased prevalence of whiteflies and cassava diseases. Sci. Rep. 10, 22049.

Kühne, T., 2009. Soil-borne viruses affecting cereals: known for long but still a threat. Virus Res. 141, 174–183.

Lapidot, M., Friedmann, M., Lachman, O., Yehezkel, A., Nahon, S., Cohen, S., Pilowsky, M., 1997. Comparison of resistance level to *Tomato yellow leaf curl virus* among commercial cultivars and breeding lines. Plant Dis. 81, 1425–1428.

Lefeuvre, P., Martin, D.P., Elena, S.F., Shepherd, D.N., Roumagnac, P., Varsani, A., 2019. Evolution and ecology of plant viruses. Nat. Rev. Microbiol. 17, 632–644.

Li, H., Roossinck, M.J., 2004. Genetic bottlenecks reduce population variation in an experimental RNA virus population. J. Virol. 78, 10582–10587.

Li, T., Zhang, J., 2001. Retroviral recombination is temperature dependent. J. Gen. Virol. 82, 1359–1364.

Li, R., Weldegergis, B.T., Li, J., Jung, C., Qu, J., Sun, Y., Qian, H., Tee, C., van Loon, J.J., Dicke, M., Chua, N.H., Liu, S.S., Ye, J., 2014. Virulence factors of geminivirus interact with MYC2 to subvert plant resistance and promote vector performance. Plant Cell 26, 4991–5008.

Lindgren, D., 1972. The temperature influence on the spontaneous mutation rate. Hereditas 70, 165–178.

Lipsitch, M., Siller, S., Nowak, M.A., 1996. The evolution of virulence in pathogens with vertical and horizontal transmission. Evolution 50, 1729–1741.

Liu, J.Y., Qian, L., Jiang, X.C., He, S.Q., Li, Z.Y., Gui, F.R., 2014. Effects of elevated CO_2 concentration on the activities of detoxifying enzymes and protective enzymes in adults of *Frankliniella occidentalis* and *Frankliniella intonsa* (Thysanoptera: Thripidae). Acta Entomol. Sin. 57, 754–761.

Lu, Z., Cui, J., Wang, L., Teng, N., Zhang, S., Lam, H.M., Zhu, Y., Xiao, S., Ke, W., Lin, J., Xu, C., Jin, B., 2021. Genome-wide DNA mutations in Arabidopsis plants after multigenerational exposure to high temperatures. Genome Biol. 22, 160.

Macfadyen, S., Paull, C., Boykin, L.M., De Barro, P., Maruthi, M.N., Otim, M., Kalyebi, A., Vassão, D.G., Sseruwagi, P., Tay, W.T., Delatte, H., Seguni, Z., Colvin, J., Omongo, C.A., 2018. Cassava whitefly, *Bemisia tabaci* (Gennadius) (Hemiptera: Aleyrodidae) in East African farming landscapes: a review of the factors determining abundance. Bull. Entomol. Res. 108, 565–582.

MacFarlane, S., Robinson, D., 2004. Transmission of plant viruses by nematodes. In: Gillespie, S., Smith, G., Osbourn, A. (Eds.), Microbe-Vector Interactions in Vector-Borne Diseases (Society for General Microbiology Symposia). Cambridge University Press, Cambridge.

Mafongoya, P., Gubba, A., Moodley, V., Chapoto, D., Kisten, L., Phophi, M., 2019. Climate change and rapidly evolving pests and diseases in Southern Africa. In: Ayuk, E., Unuigbe, N. (Eds.), New Frontiers in Natural Resources Management in Africa. Natural Resource Management and Policy, vol. 53. Springer, Cham, New York, pp. 41–57.

Makki, M., del Toro, F.J., Necira, K., Tenllado, F., Djilani-Khouadja, F., Canto, T., 2021. Differences in virulence among PVY isolates of different geographical origins when infecting an experimental host under two growing environments are not determined by HCPro. Plants 10, 1086.

Malmstrom, C.M., McCullough, A.J., Johnson, H.A., Newton, L.A., Borer, E.T., 2005. Invasive annual grasses indirectly increase virus incidence in California native perennial bunchgrasses. Oecologia 145, 153–164.

Manacorda, C.A., Gudesblat, G., Sutka, M., Alemano, S., Peluso, F., Oricchio, P., Baroli, I., Asurmendi, S., 2021. TuMV triggers stomatal closure but reduces drought tolerance in Arabidopsis. Plant Cell Environ. 44, 1399–1416.

Maree, H.J., Fox, A., Al Rwahnih, M., Boonham, N., Candresse, T., 2018. Application of HTS for routine plant virus diagnostics: state of the art and challenges. Front. Plant Sci. 9, 1082.

Martín, S., Elena, S.F., 2009. Application of game theory to the interaction between plant viruses during mixed infections. J. Gen. Virol. 90, 2815–2820.

Martínez-Arias, C., Witzell, J., Solla, A., Martin, J.A., Rodríguez-Calcerrada, J., 2022. Beneficial and pathogenic plant-microbe interactions during flooding stress. Plant Cell Environ., 1–23.

Masters, G., Brown, V., Clarke, I., Whittaker, J., Hollier, J., 1998. Direct and indirect effects of climate change on insect herbivores: *Auchenorrhyncha* (Homoptera). Ecol. Entomol. 23, 45–52.

Matsuura, S., Ishikura, S., 2014. Suppression of *Tomato mosaic virus* disease in tomato plants by deep ultraviolet irradiation using light-emitting diodes. Lett. Appl. Microbiol. 59, 457–463.

Mattenberger, F., Vila-Nistal, M., Geller, R., 2021. Increased RNA virus population diversity improves adaptability. Sci. Rep. 11, 6824.

Mauck, K.E., De Moraes, C.M., Mescher, M.C., 2010. Effects of *Cucumber mosaic virus* infection on vector and non-vector herbivores of squash. Commun. Integr. Biol. 3, 579–582.

Mauck, K., Bosque-Pérez, N.A., Eigenbrode, S.D., De Moraes, C.M., Mescher, M.C., Fox, C., 2012. Transmission mechanisms shape pathogen effects on host-vector interactions: evidence from plant viruses. Funct. Ecol. 26, 1162–1175.

Mauck, K.E., De Moraes, C., Mescher, M., 2014. *Cucumber mosaic virus*-induced changes in volatile production and plant quality: implications for disease transmission and multitrophic interactions. Phytopathology 104 (S3), 151.

Mauck, K.E., Chesnais, Q., Shapiro, L.R., 2018. Evolutionary determinants of host and vector manipulation by plant viruses. Adv. Virus Res. 101, 189–250.

Mauck, K.E., Kenney, J., Chesnais, Q., 2019. Progress and challenges in identifying molecular mechanisms underlying host and vector manipulation by plant viruses. Curr. Opin. Insect Sci. 33, 7–18.

Maule, A.J., Wang, D., 1996. Seed transmission of plant viruses: a lesson in biological complexity. Trends Microbiol. 4, 153–158.

McLeish, M.J., Fraile, A., García-Arenal, F., 2018. Ecological complexity in plant virus host range evolution. Adv. Virus Res. 101, 293–339.

Mehle, N., Guterrez-Aguirre, I., Prezelj, N., Delic, D., Vidic, U., Ravnikar, M., 2014. Survival and transmission of *Potato virus Y*, *Pepino mosaic virus* and *Potato spindle tuber viroid* in water. Appl. Environ. Microbiol. 80, 1455–1462.

Miller, Z.J., Lehnhoff, E.A., Menalled, F.D., Burrows, M., 2015. Effects of soil nitrogen and atmospheric carbon dioxide on *Wheat streak mosaic virus* and its vector (*Aceria tosichella* Kiefer). Plant Dis. 99, 1803–1807.

Mirzabaev, A., Wu, J., Evans, J., García-Oliva, F., Hussein, I.A.G., Iqbal, M.H., Kimutai, J., Knowles, T., Meza, F., Nedjraoui, D., Tena, F., Türkeş, M., Vázquez, R.J., Weltz, M., 2019. Desertification. In: Shukla, P.R., Skea, J., Calvo Buendia, E., Masson-Delmotte, V., Pörtner, H.-O., Roberts, D.C., Malley, J. (Eds.), Climate Change and Land: An IPCC Special Report on Climate Change, Desertification, Land Degradation, Sustainable Land Management, Food Security, and Greenhouse Gas Fluxes in Terrestrial Ecosystems. IPCC, Geneva, pp. 249–343.

Mitchell, C.E., Power, A.G., 2003. Release of invasive plants from fungal and viral pathogens. Nature 421, 625–627.

Mitter, N., Worrall, E.A., Robinson, K.E., Xu, Z.P., Carroll, B.J., 2017. Induction of virus resistance by exogenous application of double-stranded RNA. Curr. Opin. Virol. 26, 49–55.

Modliszewski, J.L., Wang, H., Albright, A.R., Lewis, S.M., Bennett, A.R., Huang, J., Ma, H., Wang, Y., Copenhaver, G.P., 2018. Elevated temperature increases meiotic crossover frequency via the interfering (Type I) pathway in *Arabidopsis thaliana*. PLoS Genet. 14, e1007384.

Montes, N., Pagán, I., 2019. Light intensity modulates the efficiency of virus seed transmission through modifications of plant tolerance. Plants 8, 304.

Montes, N., Vijayan, V., Pagán, I., 2021. Host population structure for tolerance determines the evolution of plant-virus interactions. New Phytol. 231, 1570–1585.

Moreno, A.B., López-Moya, J.J., 2020. When viruses play team sports: mixed infections in plants. Phytopathology 110, 29–48.

Moreno-Delafuente, A., Viñuela, E., Fereres, A., Medina, P., Trebicki, P., 2020. Simultaneous increase in CO_2 and temperature alters wheat growth and aphid performance differently depending on virus infection. Insects 11, 459.

Moreno-Delafuente, A., Viñuela, E., Fereres, A., Medina, P., Trebicki, P., 2021. Combined effects of elevated CO_2 and temperature on multitrophic interactions involving a parasitoid of plant virus vectors. BioControl 66, 307–319.

Moury, B., Fabre, F., Hébrard, E., Froissart, R., 2017. Determinants of host species range in plant viruses. J. Gen. Virol. 98, 862–873.

Moya, A., Holmes, E.C., González-Candelas, F., 2004. The population genetics and evolutionary epidemiology of RNA viruses. Annu. Rev. Microbiol. 2, 279–288.

Nachappa, P., Culkin, C.T., Saya II, P.M., Han, J., Nalam, V.J., 2016. Water stress modulates soybean aphid performance, feeding behavior, and virus transmission in soybean. Front. Plant Sci. 7, 552–567.

Nakatsukasa-Akune, M., Yamashita, K., Shimoda, Y., Uchiumi, T., Abe, M., Aoki, T., Kamizawa, A., Ayabe, S.-i., Higashi, S., Suzuki, A., 2005. Suppression of root nodule formation by artificial expression of the TrEnodDR1 (coat protein of *White clover cryptic virus 2*) gene in *Lotus japonicus*. Mol. Plant Microbe Interact. 18, 1069–1080.

Nancarrow, N., Constable, F.E., Finlay, K.J., Freeman, A.J., Rodoni, B.C., Trebicki, P., Vassiliadis, S., Yen, A.L., Luck, J.E., 2014. The effect of elevated temperature on Barley yellow dwarf virus-PAV in wheat. Virus Res. 186, 97–103.

Nebreda, M., Moreno, A., Pérez, N., Palacios, I., Seco-Fernández, V., Fereres, A., 2004. Activity of aphids associated with lettuce and broccoli in Spain and their efficiency as vectors of *Lettuce mosaic virus*. Virus Res. 100, 83–88.

Neilson, R., Boag, B., 1996. The predicted impact of possible climatic change on virus-vector nematodes in Great Britain. Eur. J. Plant Pathol. 102, 193–199.

Nifong, R.L., Gillooly, J.F., 2016. Temperature effects on virion volume and genome length in dsDNA viruses. Biol. Lett. 12, 20160023.

Nsa, I.Y., Kareem, K.T., 2015. Additive interactions of unrelated viruses in mixed infections of cowpea (*Vigna unguiculata* L. Walp). Front. Plant Sci. 6, 812.

Obrepalska-Steplowska, A., Renaut, J., Planchon, S., Przybylska, A., Wieczorek, P., Barylski, J., Palukaitis, P., 2015. Effect of temperature on the pathogenesis, accumulation of viral and satellite RNAs and on plant proteome in peanut stunt virus and satellite RNA-infected plants. Front. Plant Sci. 6, 903.

Okuda, S., Okuda, M., Matsuura, S., Okazaki, S., Iwai, H., 2013. Competence of *Frankliniella occidentalis* and *Frankliniella intonsa* strains as vectors for *Chrysanthemum stem necrosis virus*. Eur. J. Plant Pathol. 136, 355–362.

Oliveira, M.R.V., Henneberry, T.J., Anderson, P., 2001. History, current status, and collaborative research projects for *B. tabaci*. Crop Prot. 20, 709–723.

Orlov, V.N., Kust, S.V., Kalmykov, P.V., Krivosheev, V.P., Dobrov, E.N., Drachev, V.A., 1998. A comparative differential scanning calorimetric study of tobacco mosaic virus and of its coat protein *ts* mutant. FEBS Lett. 433, 307–311.

Ostfeld, R.S., Keesing, F., 2012. Effects of host diversity on infectious disease. Annu. Rev. Ecol. Evol. Syst. 43, 157–182.

Pagán, I., 2019. Movement between plants: vertical transmission. In: Palukaitis, P., García-Arenal, F. (Eds.), Cucumber Mosaic Virus. APS Press, Washington, pp. 185–198.

Pagán, I., 2022. Transmission through seeds: the unknown life of plant viruses. PLoS Pathog. 18, e1010707.

Pagán, I., García-Arenal, F., 2018. Tolerance to plant pathogens: theory and experimental evidence. Int. J. Mol. Sci. 19, 810.

Pagán, I., García-Arenal, F., 2020. Tolerance of plants to pathogens: a unifying view. Annu. Rev. Phytopathol. 58, 9.1–9.20.

Pagán, I., Alonso-Blanco, C., García-Arenal, F., 2007. The relationship of within-host multiplication and virulence in a plant-virus system. PLoS One 2, e786.

Pagán, I., Alonso-Blanco, C., García-Arenal, F., 2008. Host responses in life-history traits and tolerance to virus infection in *Arabidopsis thaliana*. PLoS Pathog. 4, e1000124.

Pagán, I., González-Jara, P., Moreno-Letelier, A., Rodelo-Urrego, M., Fraile, A., Piñero, D., García-Arenal, F., 2012. Effect of biodiversity changes in disease risk: exploring disease emergence in a plant-virus system. PLoS Pathog. 8, e1002796.

Pagán, I., Montes, N., Milgroom, M.G., García-Arenal, F., 2014. Vertical transmission selects for reduced virulence in a plant virus and for increased resistance in the host. PLoS Pathog. 10, e1004293.

Parizipour, M.H.G., Ramazani, L., Sardrood, B.P., 2018. Temperature affected transmission, symptom development and accumulation of Wheat dwarf virus. Plant Prot. Sci. 693, 222–234.

Pechinger, K., Chooi, K.M., MacDiarmid, R.M., Harper, S.J., Ziebell, H., 2019. A new era for mild strain cross-protection. Viruses 11, 670.

Peñalver-Cruz, A., Garzo, E., Prieto-Ruiz, I., Díaz-Carro, M., Winters, A., Moreno, A., Fereres, A., 2020. Feeding behavior, life history, and virus transmission ability of *Bemisia tabaci* Mediterranean species (Hemiptera: Aleyrodidae) under elevated CO_2. Insect Sci. 27, 558–570.

Pepin, K.M., Lass, S., Pulliam, J.R.C., Read, A.F., Lloyd-Smith, J.O., 2010. Identifying genetic markers of adaptation for surveillance of viral host jumps. Nat. Rev. Microbiol. 8, 802–813.

Pilet-Nayel, M.-L., Moury, B., Caffier, V., Montarry, J., Kerlan, M.-C., Fournet, S., Durel, C.-E., Delourme, R., 2017. Quantitative resistance to plant pathogens in pyramiding strategies for durable crop protection. Front. Plant Sci. 8, 1838.

Porras, M.F., Navas, C.A., Marden, J.H., Mescher, M.C., De Moraes, C.M., Pincebourde, S., Sandoval-Mojica, A., Raygoza-Garay, J.A., Holguin, G.A., Rajotte, E.G., Carlo, T.A., 2020. Enhanced heat tolerance of viral-infected aphids leads to niche expansion and reduced interspecific competition. Nat. Commun. 4, 1184.

Poudel, M., Mendes, R., Costa, L.A.S., Bueno, C.G., Meng, Y., Folimonova, S.Y., Garrett, K.A., Martins, S.J., 2021. The role of plant-associated bacteria, fungi, and viruses in drought stress mitigation. Front. Microbiol. 12, 743512.

Prasch, C.M., Sonnewald, U., 2013. Simultaneous application of heat, drought, and virus to Arabidopsis plants reveals significant shifts in signaling networks. Plant Physiol. 162, 1849–1866.

Ranabhat, N.B., Seipel, T., Lehnhoff, E.A., Miller, Z.J., Owen, K.E., Menalled, F.D., Burrows, M.E., 2018. Temperature and alternative hosts influence *Aceria tosichella* infestation and *Wheat streak mosaic virus* infection. Plant Dis. 102, 546–551.

Rentería-Canett, I., Xoconostle-Cázares, B., Ruiz-Medrano, R., Rivera-Bustamante, R.F., 2011. Geminivirus mixed infection on pepper plants: synergistic interaction between PHYVV and PepGMV. Virol. J. 8, 104.

Riedel, C., Habekuß, A., Schliephake, E., Niks, R., Broerl, I., Ordon, F., 2011. Pyramiding of Ryd2 and Ryd3 conferring tolerance to a German isolate of *Barley yellow dwarf virus*-PAV (BYDV-PAV-ASL-1) leads to quantitative resistance against this isolate. Theor. Appl. Genet. 123, 69–76.

Roberts, A.G., 2014. Plant viruses: soil-borne. In: Encyclopedia of Life Sciences. John Wiley & Sons, Hoboken.

Rochow, W.F., 1972. The role of mixed infections in the transmission of plant viruses by aphids. Annu. Rev. Phytopathol. 10, 101–124.

Rodelo-Urrego, M., Pagán, I., González-Jara, P., Betancourt, M., Moreno-Letelier, A., Ayllón, M.A., Fraile, A., Piñero, D., García-Arenal, F., 2013. Landscape heterogeneity shapes host-parasite interactions and results in apparent plant-virus codivergence. Mol. Ecol. 22, 2325–2340.

Rodelo-Urrego, M., García-Arenal, F., Pagán, I., 2015. The effect of ecosystem biodiversity on virus genetic diversity depends on virus species: a study of chiltepin-infecting begomoviruses in Mexico. Virus Evol. 1, vev004.

Román-Palacios, C., Wiens, J.J., 2020. Recent responses to climate change reveal the drivers of species extinction and survival. Proc. Natl. Acad. Sci. U. S. A. 117, 4211–4217.

Roossinck, M.J., 2012a. Persistent plant viruses: molecular hitchhikers or epigenetic elements? In: Witzany, G. (Ed.), Viruses: Essential Agents of Life. Springer, Dordrecht, pp. 177–186.

Roossinck, M.J., 2012b. Plant virus metagenomics: biodiversity and ecology. Annu. Rev. Genet. 46, 357–367.

Roossinck, M.J., 2013. Plant virus ecology. PLoS Pathog. 9, e1003304.

Roossinck, M.J., 2015. Plants, viruses and the environment: ecology and mutualism. Virology 479–480, 271–277.

Roossinck, M.J., Saha, P., Wiley, G.B., Quan, J., White, J.D., Lai, H., Chavarría, F., Shen, G., Roe, B.A., 2010. Ecogenomics: using massively parallel pyrosequencing to understand virus ecology. Mol. Ecol. 19, 81–88.

Rotenberg, D., Jacobson, A.L., Schneweis, D.J., Whitfield, A.E., 2015. Thrips transmission of tospoviruses. Curr. Opin. Virol. 15, 80–89.

Roy, B., Dubey, S., Ghosh, A., Shukla, S.M., Mandal, B., Sinha, P., 2021. Simulation of leaf curl disease dynamics in chili for strategic management options. Sci. Rep. 13 (11), 1010.
Rúa, M.A., Pollina, E.C., Power, A.G., Mitchell, C.E., 2011. The role of viruses in biological invasions: friend or foe? Curr. Opin. Virol. 1, 68–72.
Rúa, M.A., Umbanhowar, J., Hu, S., Burkey, K.O., Mitchell, C.A., 2013. Elevated CO_2 spurs reciprocal positive effects between a plant virus and an arbuscular mycorrhizal fungus. New Phytol. 199, 541–549.
Rubio, L., Herrero, J.R., Sarrió, J., Moreno, P., Guerri, J., 2003. A new approach to evaluate relative resistance and tolerance of tomato cultivars to begomoviruses causing the tomato yellow leaf curl disease in Spain. Plant Pathol. 52, 763–769.
Rubio, L., Galipienso, L., Ferriol, I., 2020. Detection of plant viruses and disease management: relevance of genetic diversity and evolution. Front. Plant Sci. 11, 1092.
Ryabov, E.V., Fraser, G., Mayo, M.A., Barker, H., Taliansky, M., 2001. Umbravirus gene expression helps *Potato leafroll virus* to invade mesophyll tissues and to be transmitted mechanically between plants. Virology 286, 363–372.
Sacristán, S., García-Arenal, F., 2008. The evolution of virulence and pathogenicity in plant pathogen populations. Mol. Plant Pathol. 9, 369–384.
Sacristán, S., Malpica, J.M., Fraile, A., García-Arenal, F., 2003. Estimation of population bottlenecks during systemic movement of *Tobacco mosaic virus* in tobacco plants. J. Virol. 77, 9906–9911.
Sacristán, S., Díaz, M., Fraile, A., García-Arenal, F., 2011. Contact transmission of *Tobacco mosaic virus*: a quantitative analysis of parameters relevant for virus evolution. J. Virol. 85, 4974–4981.
Safari, M., Ferrari, M.J., Roossinck, M.J., 2019. Manipulation of aphid behavior by a persistent plant virus. J. Virol. 93, e01781-18.
Salipante, S.J., Jerome, K.R., 2020. Digital PCR—an emerging technology with broad applications in microbiology. Clin. Chem. 66, 117–123.
Salvaudon, L., De Moraes, C.M., Mescher, M.C., 2013. Outcomes of co-infection by two potyviruses: implications for the evolution of manipulative strategies. Proc. Biol. Sci. 280, 20122959.
Santala, J., Valkonen, J.P., 2018. Sensitivity of small RNA-based detection of plant viruses. Front. Microbiol. 9, 939.
Sardanyés, J., Alcaide, C., Gómez, P., Elena, S.F., 2022. Modelling temperature-dependent dynamics of single and mixed infections in a plant virus. Appl. Math. Model. 102, 694–705.
Sastry, K.S., 2013. Seed-Borne Plant Virus Diseases. Springer, New Delhi.
Scheets, K., 1998. *Maize chlorotic mottle machlomovirus* and *Wheat streak mosaic rymovirus* concentrations increase in the synergistic disease corn lethal necrosis. Virology 242, 28–38.
Schneider, T., Kaul, C.M., Pressel, K.G., 2019. Possible climate transitions from breakup of stratocumulus decks under greenhouse warming. Nat. Geosci. 12, 163–167.
Senanayake, D.M.J.B., Mandal, B., 2014. Expression of symptoms, viral coat protein and silencing suppressor gene during mixed infection of a N-Wi strain of *Potato virus Y* and an asymptomatic strain of *Potato virus X*. Virus Dis. 25, 314–321.
Sharma, D., Yadav, U., Sharma, P., 2009. The concept of sensitivity and specificity in relation to two types of errors and its application in medical research. J. Reliab. Stat. Stud. 2, 53–58.
Shipp, J., Buitenhuis, R., Stobbs, L., Wang, K., Kim, W., Ferguson, G., 2008. Vectoring of *Pepino mosaic virus* by bumble-bees in tomato greenhouses. Ann. Appl. Biol. 153, 149–155.
Shukla, A., Pagán, I., Crevillén, P., Alonso-Blanco, C., García-Arenal, F., 2022. A role of flowering genes in the tolerance of *Arabidopsis thaliana* to *Cucumber mosaic virus*. Mol. Plant Pathol. 23, 175–187.

Simmons, H.E., Munkvold, G., 2014. Seed transmission in the *Potyviridae*. In: Gullino, M.L., Munkvold, G. (Eds.), Global Perspectives on the Health of Seeds and Plant Propagation Material. Springer, Dordrecht, pp. 3–15.

Singh, K., Wegulo, S.N., Skoracka, A., Kundu, J.K., 2018. *Wheat streak mosaic virus*: a century old virus with rising importance worldwide. Mol. Plant Pathol. 19, 2193–2206.

Singh, S., Awasthi, L.P., Jangre, A., Nirmalkar, V.K., 2020. Transmission of plant viruses through soil-inhabiting nematode vectors. In: Applied Plant Virology. Academic Press, Cambridge, pp. 291–300.

Skendžic, S., Zovko, M., Živkovic, I.P., Lešic, V., Lemic, D., 2021. The impact of climate change on agricultural insect pests. Insects 12, 440.

Sseruwagi, P., Sserubombwe, W., Legg, J., Ndunguru, J., Thresh, J., 2004. Methods of surveying the incidence and severity of cassava mosaic disease and whitefly vector populations on cassava in Africa: a review. Virus Res. 100, 129–142.

Strauss, S.Y., Agrawal, A.A., 1999. The ecology and evolution of plant tolerance to herbivory. Trends Ecol. Evol. 14, 179–185.

Stumpf, C.F., Kennedy, G.G., 2007. Effects of tomato spotted wilt virus isolates, host plants, and temperature on survival, size, and development time of *Frankliniella occidentalis*. Entomol. Exp. Appl. 123, 139–147.

Sun, Y., Guo, H., Ge, F., 2016. Plant-aphid interactions under elevated CO_2: some cues from aphid feeding behavior. Front. Plant Sci. 7, 502.

Syller, J., 2012. Facilitative and antagonistic interactions between plant viruses in mixed infections. Mol. Plant Pathol. 13, 204–216.

Szabó, A.K., Várallyay, É., Demian, E., Hegyi, A., Galbács, Z.N., Kiss, J., Bálint, J., Loxdale, H.D., Balog, A., 2020. Local aphid species infestation on invasive weeds affects virus infection of nearest crops under different management systems—a preliminary study. Front. Plant Sci. 11, 684.

Szittya, G., Silhavy, D., Molnár, A., Havelda, Z., Lovas, A., Lakatos, L., Bánfalvi, Z., Burgyán, J., 2003. Low temperature inhibits RNA silencing-mediated defence by the control of siRNA generation. EMBO J. 22, 633–640.

Taiwo, M.A., Kareem, K.T., Nsa, I.Y., D'A Hughes, J., 2007. Cowpea viruses: effect of single and mixed infections on symptomatology and virus concentration. Virol. J. 4, 95.

Tatarowska, B., Plich, J., Milczarek, D., Flis, B., 2020. Temperature-dependent resistance to *Potato virus M* in potato (*Solanum tuberosum*). Plant Pathol. 69, 1445–1452.

Tatineni, S., Graybosch, R.A., Hein, G.L., Wegulo, S.N., French, R., 2010. Wheat cultivar-specific disease synergism and alteration of virus accumulation during co-infection with *Wheat streak mosaic virus* and *Triticum mosaic virus*. Phytopathology 100, 230–238.

Tatineni, S., Wosula, E.N., Bartels, M., Hein, G.L., Graybosch, R.A., 2016. Temperature-dependent *Wsm1* and *Wsm2* gene-specific blockage of viral long-distance transport provides resistance to *Wheat streak mosaic virus* and *Triticum mosaic virus* in wheat. Mol. Plant Microbe Interact. 29, 724–738.

Tenllado, F., Díaz-Ruíz, J.R., 2001. Double-stranded RNA-mediated interference with plant virus infection. J. Virol. 75, 12288–12297.

Trębicki, P., 2020. Climate change and plant virus epidemiology. Virus Res. 286, 198059.

Trębicki, P., Nancarrow, N., Cole, E., Bosque-Pérez, N.A., Constable, F.E., Freeman, A.J., Rodoni, B., Yen, A.L., Luck, J.E., Fitzgerald, G.J., 2015. Virus disease in wheat predicted to increase with a changing climate. Glob. Chang. Biol. 21, 3511–3519.

Trębicki, P., Vandegeer, R.K., Bosque-Pérez, N.A., Powell, K.S., Dáder, B., Freeman, A.J., Yen, A.L., Fitzgerald, G.J., Luck, J.E., 2016. Virus infection mediates the effects of elevated CO_2 on plants and vectors. Sci. Rep. 6, 22785.

Trebicki, P., Nancarrow, N., Bosque-Perez, N.A., Rodoni, B., Aftab, M., Freeman, A., Yen, A., Fitzgerald, G.J., 2017. Virus incidence in wheat increases under elevated CO_2: a 4-year study of *Yellow dwarf viruses* from a free air carbon dioxide facility. Virus Res. 241, 137–144.
Truniger, V., Aranda, M.A., 2009. Recessive resistance to plant viruses. Adv. Virus Res. 75, 119–159.
Tsai, W.-A., Shafiei-Peters, J.R., Mitter, N., Dietzgen, R.G., 2022. Effects of elevated temperature on the susceptibility of capsicum plants to *Capsicum chlorosis virus* infection. Pathogens 11, 200.
Tugume, A.K., Mukasa, S.B., Valkonen, J.P.T., 2016. Mixed infections of four viruses, the incidence and phylogenetic relationships of *Sweet potato chlorotic fleck virus* (betaflexiviridae) isolates in wild species and sweetpotatoes in Uganda and evidence of distinct isolates in east Africa. PLoS One 11, e0167769.
Urban, M.C., 2015. Accelerating extinction risk from climate change. Science 348, 571–573.
Valverde, S., Vidiella, B., Montañez, R., Fraile, A., Sacristán, S., García-Arenal, F., 2020. Coexistence of nestedness and modularity in host-pathogen infection networks. Nat. Ecol. Evol. 4, 568–577.
van Munster, M., 2020. Impact of abiotic stresses on plant virus transmission by aphids. Viruses 12, 216.
van Munster, M., Yvon, M., Vile, D., Dader, B., Fereres, A., Blanc, S., 2017. Water deficit enhances the transmission of plant viruses by insect vectors. PLoS One 12, e0174398.
Van Nieuwenhove, G.A., Frías, E.A., Virla, E.G., 2016. Effects of temperature on the development, performance and fitness of the corn leafhopper *Dalbulus maidis* (DeLong) (Hemiptera: Cicadellidae): implications on its distribution under climate change. Agric. For. Entomol. 18, 1–10.
Vanitharani, R., Chellappan, P., Pita, J., Fauquet, C., 2004. Differential roles of AC2 and AC4 of cassava geminiviruses in mediating synergism and suppression of posttranscriptional gene silencing. J. Virol. 78, 9487–9498.
Vassiliadis, S., Plummer, K.M., Powell, K.S., Rochfort, S.J., 2018. Elevated CO_2 and virus infection impacts wheat and aphid metabolism. Metabolomics 14, 1–13.
Velásquez, A.C., Danve, M., Castroverde, C., He, S.Y., 2018. Plant-pathogen warfare under changing climate conditions. Curr. Biol. 28, R619–R634.
Vicedo-Cabrera, A.M., Scovronick, N., Sera, F., Royé, D., Schneider, R., Tobias, A., Astrom, C., Guo, Y., Honda, Y., Hondula, D.M., Abrutzky, R., Tong, S., de Sousa Zanotti Stagliorio Coelho, M., Saldiva, P.H.N., Lavigne, E., Correa, P.M., Ortega, N.V., Kan, H., Osorio, S., Kyselý, J., Urban, A., Orru, H., Indermitte, E., Jaakkola, J.J.K., Ryti, N., Pascal, M., Schneider, A., Katsouyanni, K., Samoli, E., Mayvaneh, F., Entezari, A., Goodman, P., Zeka, A., Michelozzi, P., de'Donato, F., Hashizume, M., Alahmad, B., Diaz, M.H., De La Cruz Valencia, C., Overcenco, A., Houthuijs, D., Ameling, C., Rao, S., Ruscio, F.D., Carrasco-Escobar, G., Seposo, X., Silva, S., Madureira, J., Holobaca, I.H., Fratianni, S., Acquaotta, F., Kim, H., Lee, W., Iniguez, C., Forsberg, B., Ragettli, M.S., Guo, Y.L.L., Chen, B.Y., Li, S., Armstrong, B., Aleman, A., Zanobetti, A., Schwartz, J., Dang, T.N., Dung, D.V., Gillett, N., Haines, A., Mengel, M., Huber, V., Gasparrini, A., 2021. The burden of heat-related mortality attributable to recent human-induced climate change. Nat. Clim. Chang. 11, 492–500.
Villamor, D.E.V., Ho, T., Al Rwahnih, M., Martin, R.R., Tzanetakis, I.E., 2019. High throughput sequencing for plant virus detection and discovery. Phytopathology 109, 717–725.
Vincent, S.J., Coutts, B.A., Jones, R.A.C., 2014. Effects of introduced and indigenous viruses on native plants: exploring their disease causing potential at the agro-ecological interface. PLoS One 9, e91224.

Wamonje, F.O., Donnelly, R., Tungadi, T.D., Murphy, A.M., Pate, A.E., Woodcock, C., Caulfield, J., Mutuku, J.M., Bruce, T.J.A., Gillgan, C.A., Pickett, J.A., Carr, J.P., 2020. Different plant viruses induce changes in feeding behavior of specialist and generalist aphids on common bean that are likely to enhance virus transmission. Front. Plant Sci. 10, 1811.

Wang, T., Keller, M.A., 2020. Larger is better in the parasitoid *Eretmocerus warrae* (Hymenoptera: Aphelinidae). Insects 11, 39.

Wang, Y., Bao, Z., Zhu, Y., Hua, J., 2009. Analysis of temperature modulation of plant defense against biotrophic microbes. Mol. Plant Microbe Interact. 22, 498–506.

Wang, G.-H., Wang, X.-X., Sun, Y.-C., Ge, F., 2014a. Impacts of elevated CO_2 on *Bemisia tabaci* infesting Bt cotton and its parasitoid *Encarsia formosa*. Entomol. Exp. Appl. 152, 228–237.

Wang, J.C., Zhang, B., Li, H.G., Wang, J.P., Zheng, C.Y., 2014b. Effects of exposure to high temperature on *Frankliniella occidentalis* (Thysanoptera: Thripidae), under arrhenotoky and sexual reproduction conditions. Fla. Entomol. 97, 504–510.

Weaver, S.C., Forrester, N.L., Liu, J., Vasilakis, N., 2021. Population bottlenecks and founder effects: implications for mosquito-borne arboviral emergence. Nat. Rev. Microbiol. 19, 184–195.

Whitham, S., McCormick, S., Baker, B., 1996. The *N* gene of tobacco confers resistance to *Tobacco* mosaic virus in transgenic tomato. Proc. Natl. Acad. Sci. U. S. A. 93, 8776–8781.

Whitlock, M.C., 1996. The Red Queen beats the jack-of-all-trades: the limitations on the evolution of phenotypic plasticity and niche breadth. Am. Nat. 148, S65–S77.

Womack, A.M., Bohannan, B.J.M., Green, J.L., 2010. Biodiversity and biogeography of the atmosphere. Philos. Trans. R. Soc. Lond. B Biol. Sci. 365, 3645–3653.

Worrall, E.A., Bravo-Cazar, A., Nilon, A.T., Fletcher, S.J., Robinson, K.E., Carr, J.P., Mitter, N., 2019. Exogenous application of RNAi-inducing double-stranded RNA inhibits aphid-mediated transmission of a plant virus. Front. Plant Sci. 10, 265.

Wosula, E.N., McMechan, A.J., Hein, G.L., 2015. The effect of temperature, relative humidity, and virus infection status on off-host survival of the wheat curl mite (Acari: Eriophyidae). J. Econ. Entomol. 108, 1545–1552.

Wosula, E.N., Tatineni, S., Wegulo, S.N., Hein, G.L., 2017. Effect of temperature on *Wheat streak mosaic* disease development in winter wheat. Plant Dis. 101, 324–330.

Wu, X., Xu, Z., Shaw, J.G., 1994. Uncoating of tobacco mosaic virus RNA in protoplasts. Virology 200, 256–262.

Wu, D., Qi, T., Li, W.X., Tian, H., Gao, H., Wang, J., Ge, J., Yao, R., Ren, C., Wang, X.B., Liu, Y., Kang, L., Ding, S.W., Xie, D., 2017. Viral effector protein manipulates host hormone signaling to attract insect vectors. Cell Res. 27, 402–415.

Wu, Y., Li, J., Liu, H., Qiao, G., Huang, X., 2020. Investigating the impact of climate warming on phenology of aphid pests in China using long-term historical data. Insects 11, 167.

Xu, P., Chen, F., Mannas, J.P., Feldman, T., Sumner, L.W., Roossinck, M.J., 2008. Virus infection improves drought tolerance. New Phytol. 180, 911–921.

Yadav, R., Chang, N.T., 2014. Effects of temperature on the development and population growth of the melon thrips, *Thrips palmi*, on eggplant, *Solanum melongena*. J. Insect Sci. 14, 78.

Yamanaka, T., Ohta, T., Takahashi, M., Meshi, T., Schmidt, R., Dean, C., Naito, S., Ishikawa, M., 2000. *TOM1*, an Arabidopsis gene required for efficient multiplication of a tobamovirus, encodes a putative transmembrane protein. Proc. Natl. Acad. Sci. U. S. A. 97, 10107–10112.

Yao, Y., Danna, C.H., Ausubel, F.M., Kovalchuk, I., 2012. Perception of volatiles produced by UVC-irradiated plants alters the response to viral infection in naïve neighboring plants. Plant Signal. Behav. 7, 741–745.

Yarwood, C.E., 1952. The phosphate effect in plant virus inoculation. Phytopathology 42, 137–143.

Ye, L., Fu, X., Ge, F., 2010. Elevated CO_2 alleviates damage from *Potato virus Y* infection in tobacco plants. Plant Sci. 179, 219–224.

Yeates, G.W., Newton, P.C.D., 2009. Long term changes in topsoil nematode populations in grazed pasture under elevated atmospheric carbon dioxide. Biol. Fertil. Soils 45, 799–808.

Yvon, M., Vile, D., Brault, V., Blanc, S., van Munster, M., 2017. Drought reduces transmission of *Turnip yellows virus*, an insect-vectored circulative virus. Virus Res. 241, 131–136.

Zhang, X., Zhang, X., Singh, J., Li, D., Qu, F., 2012. Temperature-dependent survival of *Turnip crinkle virus*-infected arabidopsis plants relies on an RNA silencing-based defense that requires dcl2, AGO2, and HEN1. J. Virol. 86, 6847–6854.

Zhao, Y., Yang, X., Zhou, G., Zhang, T., 2020. Engineering plant virus resistance: from RNA silencing to genome editing strategies. Plant Biotechnol. J. 18, 328–336.

Zhong, S.H., Liu, J.Z., Jin, H., Lin, L., Li, Q., Chen, Y., Yuan, Y.X., Wang, Z.Y., Huang, H., Qi, Y.J., Chen, X.Y., Vaucheret, H., Chory, J., Li, J., He, Z.H., 2013. Warm temperatures induce transgenerational epigenetic release of RNA silencing by inhibiting siRNA biogenesis in Arabidopsis. Proc. Natl. Acad. Sci. U. S. A. 110, 9171–9176.

Zhou, X., Harrington, R., Woiwod, I.P., Perry, J.N., Bale, J.S., Clark, S.J., 1995. Effects of temperature on aphid phenology. Glob. Chang. Biol. 1, 303–313.

Zwart, M.P., Elena, S.F., 2015. Matters of size: genetic bottlenecks in virus infection and their potential impact on evolution. Annu. Rev. Virol. 2, 161–179.

CHAPTER TWO

Marine viruses and climate change: Virioplankton, the carbon cycle, and our future ocean

Hannah Locke[a], Kay D. Bidle[b], Kimberlee Thamatrakoln[b], Christopher T. Johns[b], Juan A. Bonachela[c], Barbra D. Ferrell[a], and K. Eric Wommack[a,*]

[a]Univ. of Delaware, Delaware Biotechnology Inst., Newark, DE, United States
[b]Rutgers Univ., Dept. of Marine & Coastal Sciences, New Brunswick, NJ, United States
[c]Rutgers Univ., Dept. of Ecology, Evolution & Natural Resources, New Brunswick, NJ, United States
*Corresponding author: e-mail address: wommack@udel.edu

Contents

1. Introduction 68
2. Climate change effects on the global ocean 74
   2.1 Ocean temperature 74
   2.2 Ocean circulation 77
   2.3 Ocean stratification 77
   2.4 Ocean acidification 80
3. Key virus–host players in the marine carbon cycle 81
   3.1 Viral interactions in three bacterioplankton groups critical in the oceanic carbon cycle 82
   3.2 Viral interactions in phytoplankton groups critical in the oceanic carbon cycle 97
4. Modern approaches to investigating virus–host dynamics in a changing climate 109
   4.1 Using the meta-omics toolbox for understanding virioplankton carbon cycle dynamics 110
   4.2 Incorporating viral processes into global marine carbon cycling models 115
5. Overall takeaways and conclusions 121
Acknowledgments 122
References 122

Abstract

Interactions between marine viruses and microbes are a critical part of the oceanic carbon cycle. The impacts of virus–host interactions range from short-term disruptions in the mobility of microbial biomass carbon to higher trophic levels through cell lysis (i.e., the viral shunt) to long-term reallocation of microbial biomass carbon to the deep sea through accelerating the biological pump (i.e., the viral shuttle). The biogeochemical

Advances in Virus Research, Volume 114
ISSN 0065-3527
https://doi.org/10.1016/bs.aivir.2022.09.001

backdrop of the ocean—the physical, chemical, and biological landscape—influences the likelihood of both virus–host interactions and particle formation, and the fate and flow of carbon. As climate change reshapes the oceanic landscape through large-scale shifts in temperature, circulation, stratification, and acidification, virus-mediated carbon flux is likely to shift in response. Dynamics in the directionality and magnitude of changes in how, where, and when viruses mediate the recycling or storage of microbial biomass carbon is largely unknown. Integrating viral infection dynamics data obtained from experimental models and field systems, with particle motion microphysics and global observations of oceanic biogeochemistry, into improved ecosystem models will enable viral oceanographers to better predict the role of viruses in marine carbon cycling in the future ocean.

1. Introduction

The ocean has cumulatively absorbed ~25% of anthropogenically-released carbon since industrialization (Le Quéré et al., 2009). Much of that carbon is tied up in oceanic microbes (e.g., picoplankton, unicellular algae, and bacteria). The growth, metabolism, and death of these organisms influence the bioavailability, location, and storage of carbon. Considering that viruses can alter the carbon metabolism of their hosts (Hurwitz and U'Ren, 2016) and viral lysis results in mass mortality events contributing to the microbial loop (Fenchel, 2008; Tsai et al., 2016), virus–microbe interactions underpin and shape the flow of carbon in the ocean.

Viruses are pervasive and persistent regulators of carbon-based life and death, largely through host mortality. Lytic viruses infect a host, hijack the cell machinery for replication and virion production, and ultimately burst the infected cell, dispersing new viruses and intracellular contents into the environment. In many ways, lytic viral infection mirrors the predator–prey cycles of protistan grazers by exerting top-down control on host populations (Parsons et al., 2012; Yau et al., 2011). Moreover, viruses are not only important ecological regulators of host populations, but also lubricate biogeochemical cycling by transforming organic matter (OM). Specifically, viral lysis both facilitates carbon and nutrient exchange between trophic classes of microorganisms—autotrophs and heterotrophs—and partitions carbon and nutrients between stratified regions in the ocean.

Over the past 30 years, virus–host dynamics have been explored within specific model systems having outsized relevance for the oceanic carbon cycle. The scale of marine viral diversity, however, far outstrips any virologist's capacity to fully characterize virus–host dynamics. For example, in the Global Ocean Virome 2.0 study (Gregory et al., 2019), over 190,000 unique marine populations were identified. Metagenomics and other 'omics approaches have expanded our appreciation for the scale of marine viral diversity (Section 4.1) but are of limited use for assigning ecological function to unknown virioplankton populations. The looming challenge for viral oceanographers is connecting the vast amounts of viral genetic data coming from genomic and metagenomic studies with specific viral life history phenotypes that ultimately determine the functional impacts of virus–host interactions on the oceanic carbon cycle.

Virologists have a unique opportunity to connect infection dynamics to large-scale ecological impacts, beyond the immediate and direct regulation of host populations. While viruses can certainly be influenced by abiotic environmental factors, for example, UV (Eich et al., 2021), temperature (Kendrick et al., 2014), and pH (Fuhrmann et al., 2019), viruses are also ecosystem engineers that alter OM composition. Dissolved OM (DOM) tends to be more labile and particulate OM (POM) tends to be more recalcitrant, however, there are exceptions in both OM classes. OM transformation (DOM–POM interactions) influences the physical, chemical, and biological contexts of the ocean, which in turn, shapes whether carbon is recycled in the upper layers of the ocean (i.e., the viral shunt) or exported to deeper layers for long-term storage (i.e., the viral shuttle) (Fig. 1). The viral shunt remineralizes biomass carbon into OM supporting future rounds of carbon recycling between microbial biomass and the OM pool in the upper ocean. Each turn of the viral shunt fuels microbial respiration in the upper ocean keeping carbon available for atmospheric exchange (Bates and Mathis, 2009; Suttle, 2005). In contrast, the viral shuttle exports carbon to deeper water through processes that facilitate the sinking of OM, such as aggregation and inclusion of ballast minerals stimulated by host responses to infection. Thus, virus-released carbon or even infected cells themselves (Du Toit, 2018; Guidi et al., 2016; Laber et al., 2018; Sheyn et al., 2018) are fundamentally altered by prevailing physical and chemical conditions causing aggregation and sinking of OM to deeper oceanic layers or the benthos where carbon is sequestered from atmospheric pools for centuries to millennia (Jiao et al., 2010).

The balance of these two outcomes has profound implications for the global carbon cycle, but little is known about the interplay between the physical, biological, and chemical contexts of the ocean and how this shapes virus–host interactions to influence the fate of carbon towards the shunt or shuttle. Ocean microphysics critically impacts the probability of encounters, both between viruses and compatible hosts which results in virocells, and

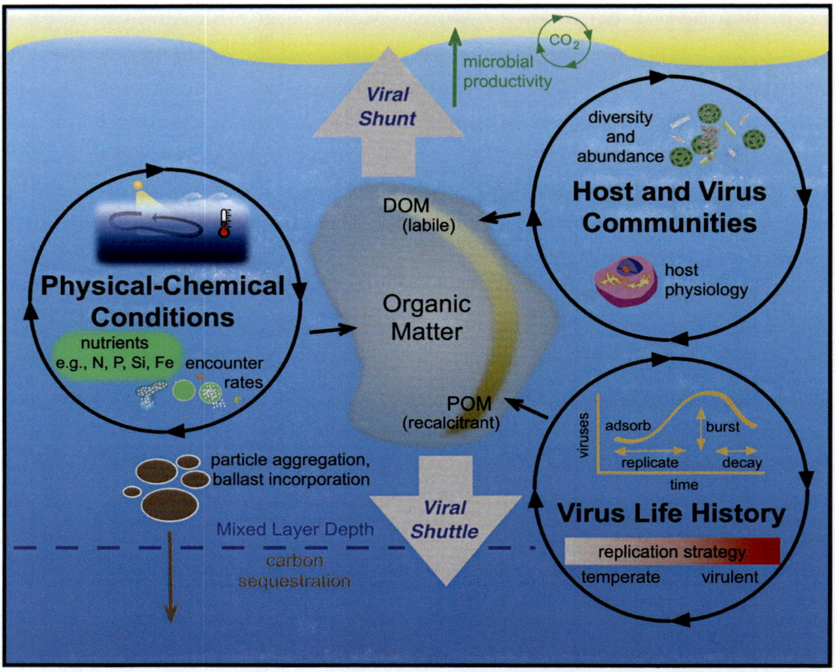

Fig. 1 Role of virioplankton in the oceanic carbon cycle. Physical–chemical conditions, community composition of resident hosts and viruses, and viral life history (temperate or virulent) influence the fate of organic matter (OM) produced from viral infection of microbial host cells. The physical and chemical contexts of the ocean are shaped by factors including temperature, light and nutrient availability, circulation, and stratification, and determine the conditions of host growth and probabilities of virus–host encounters. The diversity and abundance of host communities determine potential contributions to OM, which are influenced by host cell physiology. Ultimately, virus life history and infection impact the global oceanic carbon cycle through production of OM from the infection process. The life history of individual viral populations encapsulates the gradient of temperate to virulent replication strategy and inherent phenotypes of adsorption, lysis, and decay. It is the confluence of these physical, chemical, and biological factors that determines the chemical composition and nature of OM. Dissolved organic matter (DOM) tends to be more labile while particulate organic matter (POM) tends to be more recalcitrant, however, there are exceptions in both OM classes. Physical conditions and OM chemical composition contribute to OM transformation (DOM–POM interactions), and in turn influence whether OM produced through viral lysis is recycled and remains within surface waters (viral shunt) or is aggregated and sinks below the mixed layer (viral shuttle). The viral shunt attenuates OM into carbon dioxide through future rounds of bacterioplankton secondary production, while the viral shuttle aggregates OM with the inclusion of ballast minerals and sequesters carbon into the deep ocean. Ultimately, climate change impacts on physical and chemical conditions in the global ocean will influence the balance between viral shunt and shuttle processes and the fate of carbon within oceanic ecosystems.

between organic matter particles which results in sinking aggregates. Thus, the viral shunt and shuttle are also governed by the same encounter theory (Box 1) with importance to carbon cycling. Ultimately, it is the layering of microphysics onto the biological (e.g., host physiology, community composition) and chemical (e.g., ballast, transparent exopolymeric particles (TEP), stickiness) backdrops that shape virus replication strategy and life history. Furthermore, infection can lead to enhanced particle formation (or not), which structures the fate and flow of light-derived carbon in marine systems. To date, studies integrating these interdisciplinary factors along with diagnostic biomarkers of infection are lacking and thus limit our ability to assess viral impacts on carbon flow. Notably, these factors and their subsequent biogeochemical and ecological consequences are difficult to experimentally mimic so researchers must look to natural systems. In this review, we explore what is known for representative virus–host systems and their contribution to carbon cycling to propose a new framework of process-driven marine virology. We furthermore will argue that without advances in modeling and expansions of 'omics (Section 4), full integration of biological, chemical, and physical factors into viral frameworks will remain elusive.

While our understanding of how and when modern oceanographic processes tip the balance between viral shunt and shuttle is limited, even less is known about how climate-driven changes will impact this balance. Climate-driven changes in oceanographic processes may exacerbate, impede, or even reverse current patterns of marine virus-mediated carbon recycling and storage. Addressing these uncertainties will require physical oceanographers, biogeochemists, ecosystem modelers, and marine virologists to work together to leverage learned lessons from virus–host model systems and develop more comprehensive field and empirical approaches for meeting the challenge of understanding the role of oceanic viruses in the global carbon cycle.

Global climate change fundamentally threatens the ocean's biogeochemical balance and the interactive microbial food webs that rely on available organic and inorganic nutrients. As atmospheric carbon increases, the ocean's role as a carbon sink increases in parallel. Between 1994 and 2007, the ocean absorbed 34.4 billion tonnes of carbon alone—on top of the 118 billion tonnes of carbon already absorbed into the ocean between 1850 and 1994 (Gruber et al., 2019). Thus, the most direct effect of the ocean's absorption of atmospheric carbon (i.e., carbon dioxide (CO_2)) has been acidification which lowers cation concentration, particularly calcium,

BOX 1 Microphysical processes that shape encounters, infection, and particle formation in the oceans.

Virus–host encounters and particle formation are both governed by similar microscale physical processes but remain poorly understood. This represents a critical gap in our quantitative understanding that links predictions of infection dynamics in planktonic systems with their subsequent impact on carbon flow and marine ecosystem processes. Our understanding of natural infection processes has been limited by the neglect of biophysical mechanisms on contact between entities like viruses and their microbial hosts, the fundamental first step for infection to occur. The lytic infection rate (I) can be expressed mathematically as $I = E\delta\gamma$ where E is the rate of encounters between viruses and host cells, δ is the adsorption efficiency of a virus to host cells, and γ is the probability that a particular virus will cause lysis (infectivity). Physical encounter rates are given by $E = \beta C_1 C_2$: where β is an encounter kernel (encounters mL d^{-1}), and C_1 and C_2 are the host and virus concentrations (or particle concentrations), respectively (Burd and Jackson, 2009; Kiørboe and Saiz, 1995). Particle formation operates under the same basic principles, with a "stickiness" component (α) representing the likelihood that encounters lead to larger particle formation (Burd and Jackson, 2009) and ballast incorporation impacting a differential sinking term (see below).

Viruses, most known phytoplankton hosts, and particles are non-motile, with encounters depending on other physical processes such as random diffusion (Brownian motion), differential sinking, and turbulence. Hence, three potential encounter mechanisms can be considered: Brownian motion (β_M), differential sinking (β_S), and turbulent water motion (β_T). Rates of encounter by these mechanisms vary with the size and density of individual particles. β can be expressed as a sum of the individual encounter kernel expressions (Burd and Jackson, 2009; Kiørboe and Saiz, 1995) and this framework can be used with empirical data to quantify the infection rate of different cell systems and associated particle formation across different ocean regions (K. Bondoc, Personal communication, 2022). Studies of viral infection typically consider only Brownian motion (Brown and Bidle, 2014; Cottrell and Suttle, 1995b; Johns et al., 2019; Nissimov et al., 2019a, b), but encounter rates can be higher due to sinking, especially for ballasted cells, and due to fluid motion (i.e., turbulence). Host cells (such as ballasted phytoplankton like coccolithophores and diatoms) also have relatively high sinking rates due to their calcium carbonate and biogenic silica mineral ballast, but the role of differential sinking has been neglected in virus ecology. Likewise, micro-scale turbulence facilitates encounters by increasing the relative speeds of particles, but this mechanism has also largely been ignored (Basterretxea et al., 2020). Turbulent kinetic energy (TKE) is injected into the ocean via physical processes including wind stress and forms a cascade of eddies from scales of meters to 10s or 100s of meters, down to the smallest, Kolmogorov-scales (<1 cm), where the energy dissipates (Margalef, 1998).

BOX 1 Microphysical processes that shape encounters, infection, and particle formation in the oceans.—cont'd

The dissipation rate (ε) of TKE varies over many orders of magnitude (Franks et al., 2022; Fuchs and Gerbi, 2016) and we expect it to drive considerable environmental variation in infection rates. The underlying principles of microscale ocean physics shape the integrated outcomes of virus infection, whether it leads to particle formation, and the degree to which infection is coupled to shunt versus shuttle.

negatively impacting calcifying marine organisms including microbes (Doney et al., 2009; Kroeker et al., 2013; Yamamoto-Kawai et al., 2009). However, of equally important concern for oceanic ecosystems are the broader climatic changes resulting from intensification of the greenhouse effect due to increasing levels of atmospheric warming gasses (e.g., CO_2, methane (CH_4)). These impacts include increasing sea surface temperature (Seager et al., 2019) and stratification-induced changes in ocean circulation (Li et al., 2020), which have profound implications for many oceanic biogeochemical processes (Section 2). Like all other marine organisms, microbes are sensitive to climate-induced changes in circulation, nutrient availability, and pH, which likely means that changes in biogeochemical or oceanographic context will have cascading effects. Ocean circulation and stratification also shape the productivity of oceanic food webs where upwellings of cold, nutrient-rich waters spur phytoplankton blooms, consequently shaping the migration and population dynamics of microzooplankton grazers (Batchelder et al., 2002; Edwards et al., 2000; Neuer and Cowles, 1994; Smayda, 2010), as well as macroorganisms such as fish and whales (Croll et al., 2005). The difficulty in understanding how physical and chemical conditions modified by global climate change will impact virus–host interactions and consequently virus-mediated marine processes mostly lies in the difficulty of scale (from single interacting virus–host populations to ocean-scale ecosystems).

Given the urgency to understand how climate change will fundamentally alter the processes that shape the directionality and efficiency of virus-mediated carbon cycling, this review will integrate knowledge of widely varied concerns in a holistic approach aimed towards understanding how viral processes may change the trajectory of oceanic carbon cycling as the ocean responds to global climate change. Specifically, we (1) briefly review the current framework for the factors that tilt the microbial loop towards either shunt- or shuttle-domination, (2) explore how climate-induced changes to the underlying oceanographic processes could skew the composition, amount, and fate of lysis-derived organic materials, (3) highlight the key viral players that, to our current understanding, are critical to carbon cycling responses to climate change, and (4) outline the critical role that modern 'omics approaches and modeling plays in closing the gap between the data generated by experimentalists and the nuanced complexities of the ecological totality.

2. Climate change effects on the global ocean

Unprecedented atmospheric and oceanic CO_2 inputs are already altering fundamental abiotic conditions in marine systems, specifically patterns of temperature, circulation, stratification, and ocean pH (Fig. 2). Here we briefly review how these four oceanic features are critical to marine life and how each is responding to global climate change.

2.1 Ocean temperature

The ocean absorbed 93% of excess heat produced by global climate change between the 1970s and 2010 (Stocker, 2014). Consequently, the ocean is warming continuously, albeit unevenly across depths. The upper layer (epi- and meso-pelagic between 0 and 700 m where mixing principally occurs) has warmed at a rate of 5.31 ± 0.48 zettajoule (ZJ, or 10^{21} J) yr^{-1} (Bindoff et al., 2019). In contrast, the lower layer between 700 and 2000 m (roughly bathypelagic and lower) has warmed at a slower rate of 4.02 ± 0.97 ZJ yr^{-1}. To place these rates into understandable context, humans consume approximately 0.5 ZJ yr^{-1} (Tomabechi, 2010), although this energy demand is continuously growing.

Increasing average water temperature disrupts thermohaline gradients that drive ocean circulation (Section 2.2) and stratification (Section 2.3), but there are also often overlooked biological effects of temperature. Water temperature can directly alter host physiology, and indirectly alter the probability and dynamics of infections. For example, lytic cyanophage favor shortened latent periods and increased burst sizes under warmer conditions (Steenhauer et al., 2016; Yadav and Ahn, 2021). In contrast, warming confers improved resistance in *Emiliania huxleyi* against lytic infections, although the mechanism underpinning this response is unclear (Kendrick et al., 2014).

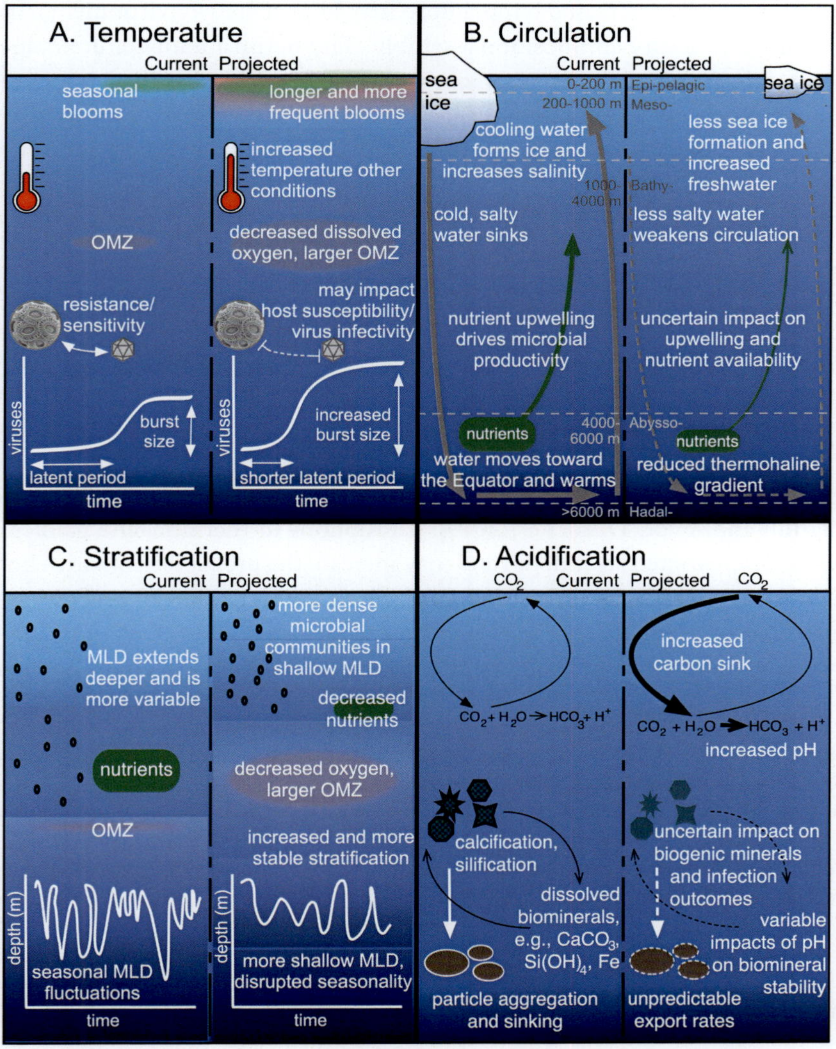

Fig. 2 Climate change effects on the global ocean. Climate change is fundamentally altering patterns of (A) temperature, (B) circulation, (C) stratification, and (D) acidification. (A) Increased temperatures tend to boost microbial productivity, suggesting increased frequency and duration of bloom events. Warmer water holds less dissolved oxygen, contributing to the expansion of oxygen minimum zones (OMZ). Increased temperatures also shift outcomes of viral infection, influencing host susceptibility and/or virus infectivity, as well as shortening the latent period and increasing burst size. Contributing to sea ice melt and salinity, temperature is intimately linked with circulation and stratification. (B) Thermohaline gradients drive global ocean currents. Cold, salty water at polar surface depths sinks and travels toward the Equator.

(Continued)

Additionally, increased temperature may impose new environmental filters on community composition assembly. The optimal temperature ranges of cyanobacteria and green algae overlap, and surpass the optimal ranges for both dinoflagellates and diatoms (Paerl, 2014). Consequently, warmer temperatures may tip the scales in favor of cyanobacteria and green algae. At the very least, warming waters will promote range expansion of warm-tolerant species in replacement of cold-evolved species (Edwards and Richardson, 2004; Richardson and Schoeman, 2004). Despite these generalizable patterns, predicting ecological community succession with warming marine waters remains difficult at best due to intraspecific genetic diversity within phytoplankton populations and species-specific differences between populations. For example, while modeling predicts warmer oceans will lead to earlier bloom phenology in the spring as well as more frequent bloom events (Mészáros et al., 2021), phytoplankton demonstrate wide species-specific responses to warmer waters ranging from tolerant to sensitive (Barton and Yvon-Durocher, 2019). Disruptions to thermohaline gradients will alter the mixed layer depth and may increase the probability of virus–host interactions through increasing particle concentration (see Box 1). Importantly, an increase in average water temperature, as well as an increase in frequency and intensity of marine heat waves, may act as an environmental

Fig. 2—Cont'd The warming water is pushed toward the surface, creating upwelling zones that bring nutrients to the surface and support microbial productivity. Climate change will disturb current circulation patterns primarily through increased sea ice melt and reduced sea ice formation. Changes in upwelling and nutrient availability will impact microbial productivity, though their effects are uncertain. (C) Vertical stratification is primarily defined by water density gradients (impacted by thermohaline gradients). Stratification sets a physical boundary that defines the mixed layer depth (MLD) that varies seasonally, across ocean regions, and in response to turbulence. Global climate change is predicted to make the MLD shallower and more stable, increasing the impacts of characteristic increased microbial density and decreased nutrient and oxygen availability on microbial communities. (D) The delicate balance of atmospheric carbon dioxide (CO_2) exchange maintains seawater pH and releases carbonic acid with cascading effects on biomineral availability and stability. Mineral chemistry impacts particle composition and sinking, facilitating carbon export to deeper waters. Increasing CO_2 production disrupts this balance, reducing calcium carbonate or silica availability that is critical for marine calcifying organisms (e.g., coccolithophores) and diatoms, respectively. In turn, this will impact particle formation and carbon export. Variations in bioavailability of typically limited nutrients (e.g., iron) have anticipated effects on microbial productivity. The cumulative effect on temperature, circulation, stratification, and acidification by global climate change will profoundly alter microbial communities with relevance to carbon flow via the viral shunt or shuttle.

filter or a narrow bottlenecking event that alters phytoplankton community composition and diversity (Lindh et al., 2013; Striebel et al., 2016; Thomas et al., 2016; Vallina et al., 2017).

2.2 Ocean circulation

Thermohaline circulation drives ocean currents through differences in seawater density. As colder, high salinity water sinks, it is replaced by warmer, less saline waters, effectively creating a global conveyor belt of latitudinal water movement around the Atlantic, Pacific, Indian, and Southern Oceans. Sinking polar waters move down vertically and migrate latitudinally along the ocean bottom toward tropical latitudes. As cold deep water returns to the warmer tropics, it is forcibly moved up by the continuous pushing of water behind. In these upwelling zones, cold deep waters bring nutrients to the surface stimulating microbial food webs (Armengol et al., 2019; Vargas et al., 2007).

Climate change decelerates existing circulation patterns primarily through the destabilization and reduction of sea ice formation (Silvano et al., 2018). Increased freshwater inputs in arctic regions from ice melt and increased precipitation decrease the density gradient traditionally produced by strong thermohaline differences (Carlson and Clark, 2012; Farmer et al., 2021; Zaucker et al., 1994). The Atlantic Meridional Overturning Circulation (AMOC) is a well-studied system responsible for distributing warm, tropical water to northern latitudes where it cools and sinks, pushing back down to the tropics continuing the cycle. Currently, the AMOC is moving at its slowest rate in approximately 1000 years (Caesar et al., 2020). Likewise, models predict slowdowns in other regional circulations, such as the Southern Indian Ocean (Stellema et al., 2019).

2.3 Ocean stratification

The strength of vertical stratification is defined by the degree of density difference between warmer waters near the top of the water column and colder waters below. Climate change may alter natural stratification patterns by making regions of the ocean both more stratified and more stable, specifically altering the mixed layer depth (MLD) (Box 2). Naturally, the degree of stratification differs across global ocean regions, seasonal changes, and even in response to episodic storms (Diaz et al., 2021). Increasing surface water temperatures may exacerbate existing seasonal extremes by creating shallow MLDs during stretches of warm weather, or, alternatively, could

BOX 2 Consequences of the mixed layer depth on oceanic carbon cycling.

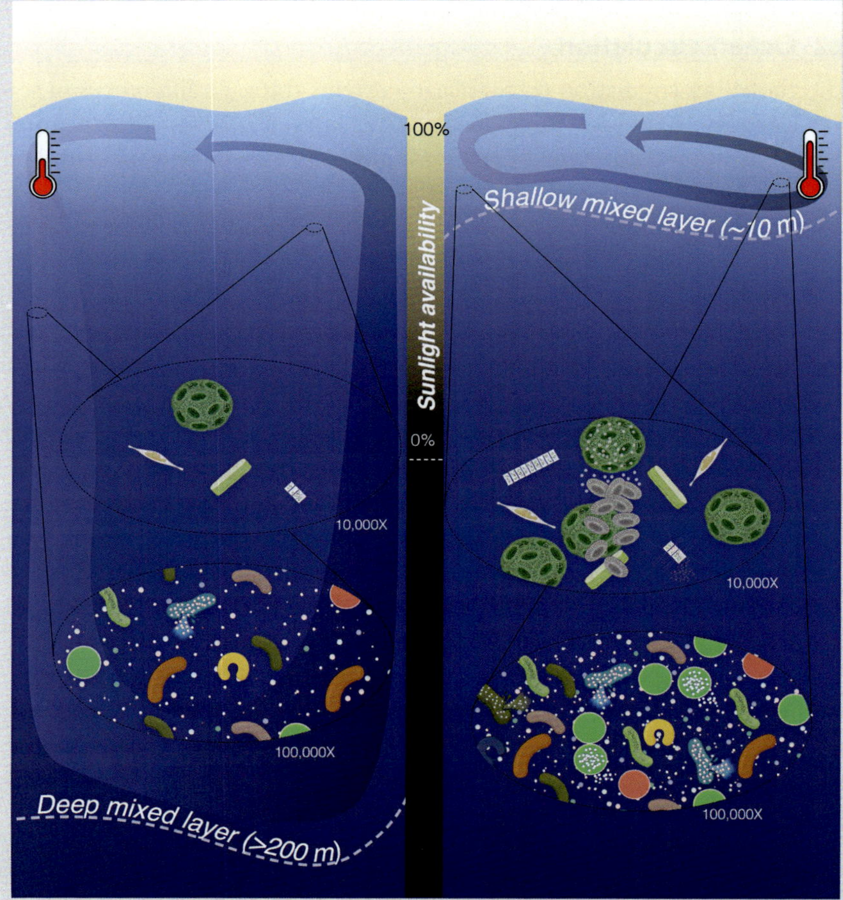

The surface mixed layer (ML) represents the uppermost region in the ocean that is homogenized by turbulent mixing or convective overturning. The physical boundary and plasticity of the mixed layer depth (MLD) has profound consequences on marine microbial ecology, microbial productivity, and virus–host interactions, and thus oceanic carbon availability and cycling. Deep ML (left panel) are characterized by cooler surface waters, weaker stratification, and water mass movement that distribute microbial communities well below the euphotic zone (indicated by yellow/black gradient shading in middle bar). Here, mixing occurs over a larger volume, which effectively distributes and dilutes microbial communities and causes photoautotrophs to spend less time within the euphotic zone, thereby imposing light limitation and reducing net primary production (PP)

BOX 2 Consequences of the mixed layer depth on oceanic carbon cycling.—cont'd

of the system. Reductions in PP cascade through the community reducing secondary (bacterioplankton) production, which is dependent on organic carbon in the system. Dilution of hosts along with lowered host physiological activity lead to less viral production and a general slowing of virus-mediated carbon release. These conditions would favor viruses having temperate or pseudotemperate life cycles. In contrast, shallow ML (right panel) are characterized by stronger stratification, higher integrated light levels through the day, warmer surface waters, and little to no mass water movement below the euphotic zone. Here, mixing occurs over a smaller volume, effectively concentrating microbial communities in near surface waters and providing conditions whereby photoautotrophs spend more time within the euphotic zone, maximizing net PP fueling increases in secondary production. Shallow ML conditions lead to increases in host growth and virus–host encounter rates that would favor lytic viral life cycles, enhanced virus production, and organic carbon release.

Seasonal phytoplankton blooms are intimately linked with MLD. As daily temperatures and sunlight increase through the spring, the MLD shallows, exposing phytoplankton to higher, integrated daily irradiance, resulting in increased system productivity, high surface biomass accumulation, and an injection of organic carbon into oceanic ecosystems (Behrenfeld and Boss, 2018; Smith et al., 2015). Indeed, varied ocean systems, perhaps epitomized by those found in the North Atlantic, host large seasonal blooms, whose phases and planktonic productivity are shaped by MLD over days to months (Behrenfeld et al., 2019; Bolaños et al., 2021; Diaz et al., 2021; Fox et al., 2020; Graff and Behrenfeld, 2018; Morison et al., 2019; Penta et al., 2021). Upon subsequent shallowing, phytoplankton communities rapidly grow and accumulate biomass but exhibit pronounced signatures of oxidative stress, transparent exopolymer particle production, and positive viral production (Diaz et al., 2021). Prolonged stratification (like that seen in summer and fall) leads to nutrient deprivation in the ML, decreases in phytoplankton concentrations, negative particulate accumulation rates, signatures of compromised membranes, death-related protease activities, and virus production (Diaz et al., 2021). Thus, the seasonal variation of MLD sets the biophysical parameters and pace of photo-derived carbon into the microbial food web. Storms can episodically disturb seasonal stratification, deepening the ML to 200 m or greater, a depth well below the euphotic zone, replenishing the ML with nutrients from below but accompanied with transient light-limitation of phytoplankton (Diaz et al., 2021; B. Diaz, Personal communications, 2022). Subsequent stratification (over several days) can then reset virus–host dynamics. Given these observations, projected changes in MLD, namely enhanced stratification (Li et al., 2020) resulting from global climate change, may critically impact virus–host interactions and carbon cycle outcomes in oceanic microbial communities.

disrupt natural patterns by creating a long-term or semi-permanent shallow MLD. Since the 1960s, ocean stratification has increased at a rate of 0.9% each decade, yielding an astonishing total change of 5.3% increased stratification by 2018 (Li et al., 2020). Increases in temperature have driven more than 90% of this stratification change over the last 60 years, with greater than 70% change accounted for by increased warming in the upper layers (Li et al., 2020).

The ecological and biogeochemical consequences of increased ocean stratification are numerous. Intergovernmental Panel on Climate Change modeling predicts that increases in stratification will reduce nutrient availability in the upper layers where most of the ocean's biological productivity occurs. In the tropics alone, nutrient availability will shrink by 7–16% by 2081–2100 (Cassotta et al., 2022). Intensified stratification also reduces oxygen availability; between 1970 and 2010, oxygen availability above 200 m shrank by 0.5–3%, whereas oxygen minimum zones occurring at mesopelagic depths grew by 3–8% (Bindoff et al., 2019). All of these effects serve to reduce the biological productivity of the surface ocean where atmospheric carbon is transformed into microbial biomass.

2.4 Ocean acidification

Throughout global industrialization, the ocean has acted as a major carbon sink, absorbing approximately 30% of all CO_2 emissions since 1750 (Guinotte and Fabry, 2008). However, ocean acidification is the cost of the ocean's buffering against CO_2 concentrations greater than the current 420 ppm. Since the start of the Industrial Revolution, the average pH of the ocean has dropped by 0.1 pH units, with an expected further drop of 0.3 pH units by the end of the 21st century (Feely et al., 2009; Guinotte and Fabry, 2008). The consequences of this phenomena are numerous—ocean acidification alters the chemistry of seawater in marine food webs, as the release of carbonic acid into the ocean creates cascading effects on the bioavailability and dissolution of minerals. Hence, ocean acidification will have contrasting effects on biomineral production and dissolution, depending on mineral chemistry, which will subsequently impact particle composition. The relationship between acidification, virus infection, and the facilitation (or not) of sinking will depend on the relative balance between biomineral ballast production and dissolution (preservation). As pH impacts biomineral stability, it stands to reason that fewer biominerals will be incorporated into particles which would subsequently reduce sinking of carbon captured within particles. Famously, ocean acidification erodes the stability of biogenic calcium carbonate minerals which is critical for marine calcifying organisms, such as coccolithophores (Riebesell and Gattuso, 2015; Riebesell et al., 2017). Furthermore, coccolithophores' calcifying ability has key implications for virus infection outcomes, given that an intact coccosphere affords the cell some protection against infection (Johns et al., 2019) and that free-floating planktonic calcite biominerals can also adsorb and deliver viruses to cells with greater efficiency than free viruses on their own (Johns et al., 2019; C.T. Johns, Personal communication, 2022). Ocean acidification has also been found to impair diatom silica production (Petrou et al., 2019) which could mean less silica ballast would be available for particle incorporation. However, given potential decreases in biogenic silica dissolution at low pH, preservation of biogenic silica could be enhanced by ocean acidification leading to enhanced ballast-mediated export (Taucher et al., 2022).

Ocean acidification can also impact the availability of critical nutrients. For example, low pH increases the bioavailability of the limiting trace metal iron (Gledhill et al., 2015) so it stands to reason that microbial systems may be freed from iron limitation and may experience enhanced productivity. Taken together, the physical and chemical changes occurring to the global ocean as a result of climate change will profoundly alter the ocean's biological systems in ways that are poorly understood by scientists and with relevance to carbon flow through the viral shunt and shuttle pathways.

3. Key virus–host players in the marine carbon cycle

Ever since the adoption of the philosophical framework of the microbial loop (Pomeroy, 1974), microbial ecologists have focused on unraveling its underlying mechanistic details. Decades of investigation, leveraging increasingly sensitive molecular genetic tools and DNA sequencing, have unveiled the composition and diversity of marine microbial communities to a point where some microbial taxa can be unambiguously assigned specific biogeochemical roles within the oceanic carbon cycle. In the following subsections of this review, we highlight the ecological and biological features of known viruses infecting some of the key microbial groups known to have outsized impacts on the marine carbon cycle. In particular, we examine virus–host interactions and their carbon cycle implications within key microbial host groups: three bacterioplankton host groups—heterotrophs *Pelagibacter* and *Roseobacter*, and the obligate photoautotrophs, cyanobacteria—and multiple microalgal phytoplankton host groups—haptophytes, chlorophytes, diatoms, and dinoflagellates. While not an exhaustive summary of all known ecologically important viruses of microbes, these sections highlight key themes and research questions applicable to understanding the implications of viral infection for specific microbial taxa shaping the flow of carbon and nutrients within marine ecosystems.

3.1 Viral interactions in three bacterioplankton groups critical in the oceanic carbon cycle

Starting with the discovery of the great plate count anomaly (MacLeod, 1965), where the number of observable host cells exceeds the number of cultivable host cells by 10-fold or more, microbial ecologists began the continuing intellectual process of asking three fundamental questions essential in dissecting the black box of microbial communities: Who's there? How many of each are there? What are they doing? Succeeding rounds of technological advancement have improved the sensitivity and precision of answers to these questions. With each round, studies of marine microbial communities have led the way owing to both the importance of oceanic ecosystems and the ease of obtaining relatively unadulterated samples of microbial cells in different size classes by filtration (Mueller et al., 2014), and viruses by filtration (Wommack et al., 2010) or chemical flocculation (John et al., 2011). Early application of molecular cloning and sequencing of small subunit rRNA genes (16S rDNA) PCR-amplified from marine microbial communities uncovered the pelagic ocean's vast bacterial diversity. Here we focus on three dominant bacterioplankton groups with outsized contributions to carbon and nutrient cycling.

3.1.1 Pelagiphages

Early evaluations of oceanic microbial diversity revealed a particular predominant but unknown 16S clade, initially termed clade SAR11 (after the Sargasso Sea where it was first isolated) and later assigned to the order Pelagibacterales (Giovannoni, 2017; Giovannoni et al., 1990). Observations collected using a variety of molecular genetics and microscopy approaches indicate that members of the SAR11 clade are nearly ubiquitous in the world ocean occurring from the mesopelagic zone to the surface. These important microbes reach their highest abundances in the euphotic zone of the pelagic ocean oligotrophic gyres (Morris et al., 2002), systems of rotating ocean currents and the world's largest ecosystems by volume and area (Karl, 1999). Given this distribution, SAR11 accounts for 25% of all plankton with an estimated global abundance of 2.4×10^{24} cells (Giovannoni, 2017). SAR11's dominance indicates that it is a critically important player in the oceanic carbon cycle where its niche is as an oligotroph capable of assimilating a wide range of simple growth substrates. In pelagic ecosystems, SAR11 populations are responsible for assimilating half of available amino acids and 30% of dimethylsulfoniopropionate (DMSP), a common osmolyte of microalgae (Malmstrom et al., 2004), as well as a variety of C1 compounds (Sun et al., 2011) including carbon monoxide (CO), CO_2, and CH_4. The fact that SAR11 utilizes DMSP as a growth substrate and produces dimethylsulfide (DMS) gas is noteworthy, as DMS plays an important role in climate as a cloud condensation molecule (Sun et al., 2016).

While *in situ* evidence of a bacteria's presence and abundance informs ecological investigation and genomic evidence provides clues as to its potential physiology, only laboratory cultivation can provide purified virus–host systems and, thus, model experimental systems for reductionist investigations. As could be hypothesized for a bacterial specialist living in a nutrient-limited oligotrophic environment, SAR11 cells were among the smallest cells ever observed having a high surface area to volume ratio (vibroid shape, 0.37–0.89 μm in length, and 0.12–0.2 μm in diameter). Under nutrient-replete conditions SAR11 cultures reach maximal densities of $\sim 3.5 \times 10^6$ cells mL^{-1}, with doubling times around 0.5 d^{-1}, which compares well with *in situ* rates measured for SAR11 populations (Malmstrom et al., 2005) and slightly greater than the high end range ($\sim 0.3\ d^{-1}$) observed for bacterioplankton communities within the oligotrophic gyres (Rappé et al., 2002; Schwalbach et al., 2010). The phenotypic features of cultivated SAR11 strains, which include slow growth, low maximal cell density, and diminutive cell size, contrast greatly with copiotrophic bacterial pathogens such as *Vibrio*, *Pseudomonas*, *Salmonella*, *Escherichia*, and others that are more commonly studied hosts for virus–host experimental models. Given the unique challenges of growing SAR11 in the lab, it took more than a decade before the first isolated SAR11 viruses were reported (Rappé et al., 2002; Zhao et al., 2013). The lack of evidence for viruses infecting SAR11 during the decade between the first isolation of SAR11 hosts and the first isolation of its phages fueled hypotheses that SAR11 was an exquisite defense specialist somehow immune to widespread viral infection. This notion of defense specialism was brought into question by the discovery of SAR11 viruses, termed pelagiphages (a nod to the host genera), and remains an intriguing debate (Våge et al., 2013).

To date, 46 pelagiphages have been isolated by dilution-to-extinction cultivation approaches on both cold and warm strains in SAR11 subclade 1a. Most of these phages have been isolated on the cold strains *Candidatus Pelagibacter ubique* type strain HTCC1062 and strain FZCC0015 from the

Taiwan strait, with additional phages isolated on the warm strains *P. bermudensis* HTCC7211 and H2P3α from the Western English Channel (Buchholz et al., 2021a, b; Du et al., 2021; Zhang et al., 2019b, 2021; Zhao et al., 2013). While many details of the interactions between pelagiphages and their ubiquitous SAR11 hosts remain a mystery, thanks to metagenomics we know a surprising amount about the biogeographic distribution of these phages. Bioinformatically "mapping" DNA sequence reads from oceanic virome libraries (Section 4.1) against newly discovered pelagiphage genomes revealed that these phages recruited an astounding 60% of the assignable reads from the Pacific Ocean Virome dataset (Hurwitz and Sullivan, 2013) in analyses containing other phages existing within marine ecosystems (Zhao et al., 2013). Subsequent studies using even larger collections of virome sequence libraries from across the global ocean and more pelagiphages have also shown the high frequency of pelagiphages in both coastal and pelagic ocean environments (Buchholz et al., 2021a, b; Zhang et al., 2021). Reflecting their host's distribution, pelagiphage abundance is highest in the epipelagic but is still notable for some phages in the deeper waters of the meso- and bathypelagic (Zhang et al., 2021). Despite the modest number of cultivated pelagiphages it is already clear that pelagiphages demonstrate dramatic differences in their abundance. Some pelagiphages, such as those isolated from the Western English Channel (Buchholz et al., 2021a, b), appear to be rare within the virioplankton, recruiting few virome reads and supporting prior observations of the long tail of rare viral populations within the virioplankton (Breitbart et al., 2018). In contrast, every pelagiphage sequence mapping experiment has demonstrated that HTCV010P, a podovirus, is among the most commonly observed viruses throughout the global ocean (Buchholz et al., 2021a, b; Kang et al., 2013; Zhang et al., 2021; Zhao et al., 2013). Despite its abundance, HTCV010P is the most unknown of all cultivated pelagiphages. Only seven of its 64 predicted open reading frames demonstrated homology to a functionally annotated protein. Besides a peptidase and the large and small subunit terminase genes, all are structural proteins. The fact that arguably the ocean's most abundant and ubiquitous virus is a genetic enigma is both exciting and humbling. However, unwrapping the biology of this virus will undoubtedly aid efforts towards understanding the role of pelagiphages in the oceanic carbon cycle.

The majority of cultivated pelagiphages (42) have a podovirus morphology with a short, non-contractile tail, and genome sizes ranging from 31.6 to 60.9 kb. Host specificity is the norm among pelagipodophages, although a broad host range was observed for four isolated phages infecting both cold and warm SAR11 ecotypes (Zhao et al., 2019a) and three infecting two or three warm ecotype strains (Buchholz et al., 2021a, b). The two reported pelagiphages with a myoviral morphology, HTCV008M (Zhao et al., 2013) and EXVC030M Mosig (Buchholz et al., 2021a, b), have long contractile tails, larger capsids, and substantially larger genomes (141 and 147 kb, respectively) than the pelagipodophages. Lastly, two pelagiphages having siphoviral morphology have been reported, EXVC016S Kólga and EXVC013S Aegir (Buchholz et al., 2021a, b), with long non-contractile tails and genomes more similar in size to podoviruses (EXVC016S Kólga, 48.7 kb; EXVC013S Aegir genome size unknown).

All pelagiphage isolation and cultivation has been done in liquid cultures as SAR11 cannot be grown on solid media. Because of this constraint, plaque assay is not possible with currently known pelagiphages and all assessments of virus–host dynamics have been inferred from direct microscopic observations of phage particles and host cells. As a consequence, the virus–host infection dynamics of only four phages have been characterized (Zhao et al., 2013), while the genomes of all 46 have been reported. The infection dynamics of three pelagiphages belonging to the podoviridae morphological group (HTVC011P, 019P, and 010P) were remarkably similar, with burst sizes 37–49 viruses per cell lysed and a latent period of 19–24 h before the first increases in viral count. In contrast, the myovirus HTCV008M demonstrated a low burst size of nine viruses and shorter latent period of 16–19 h. These latent periods fit with reported *Pelagibacter* biological characteristics in that phage lysis timing is modestly longer than the average reported host doubling time, and burst sizes reflect that the net amount of dsDNA within the released phage particles is comparable to the size of a typical *Pelagibacter* genome (~1.5 Mb). Limited availability of deoxyribonucleotides for phage genome replication could be particularly acute for the larger genome of the myovirus HTCV008M (147 kb genome) and may explain the unusually low burst size for this phage. After three to four infection rounds (60–70 h), pelagiphage abundances exceeded *Pelagibacter* host cell abundance by 5 to 10-fold with host cell abundance ~10-fold lower than uninfected control cultures. Although not discussed by the authors (Zhao et al., 2013), there were notable differences in the loss of host cells between the phages, with podoviruses HTCV011P and HTCV019P showing a more rapid rate of host cell loss and 5-fold lower host abundance at 60 h than podovirus HTCV010P and myovirus HTCV008M. Gene content variation could be linked with these observed differences in virus:host ratios and host cell death rates. For instance, 011P and O19, both with observed high virus:host ratios and increased host cell loss, carry genes encoding RNA polymerase, DNA replication proteins (Family A DNA polymerase (PolA), primase), integrase, and lysozyme. In contrast, 010P carries none of these genes. HTVC008M also does not carry an RNA polymerase and utilizes a different family B DNA polymerase (PolB) that is more commonly seen in myoviruses.

Because of the lack of virus–host infection observations, most of what we know of the pelagiphages comes from their genome sequences. All genomes have low GC content (32.6% ± 1.4%), similar to that of SAR11 and indicative of organisms thriving under nitrogen-limiting conditions (Grzymski and Dussaq, 2012). No single gene is universally conserved within this group and on average only 34% of predicted ORFs showed homology to a known gene (Buchholz et al., 2021a,b; Zhao et al., 2013). Nevertheless, there are some important functional genes that are shared among the group. In particular, each of the known pelagiphages carries at least one or more of an important group of four genes—terminase, DNA polymerase (DNAP), RNA polymerase, or ribonucleotide reductase (RNR)—that enable examination of deep phylogenetic relationships with other viruses and cellular life and can inform hypotheses concerning the physiological and genetic constraints over viral replication. In particular, gene content recapitulates phylogenetic relationships, often separating pelagiphage into distinct clades that reflect morphology. The DNA packaging protein, terminase, clearly separates the podophages into three clades separate from two other clades each containing a known siphovirus and from the two myoviruses (Buchholz et al., 2021a, b). Sixty percent of the pelagiphages carry a DNAP, with the two myoviruses carrying PolB and 27 podoviruses carrying PolA. All DNAP-carrying pelagiphages also carry a primase suggesting that their genome-encoded polymerases are principally responsible for viral genome replication. Among the DNAP-carrying pelagiphages, all but nine also carry RNR indicating that these phages have the capability of controlling the supply of deoxyribonucleotides available for genome replication. As observed in other genes, the RNR carried by myoviruses differed from podoviruses, existing in distinct phylogenetic clades (Class Ia subclass NrdAg and NrdAk, respectively).

An additional functional gene indicative of potential pelagiphage infection dynamics is integrase, carried in 41% of the genomes. Discovery of integrase within so many cultivated pelagiphages was somewhat surprising, because all of these phages propagate via lytic infection, and subsequent metagenomic read mapping analysis (Zhang et al., 2021), qPCR (Eggleston and Hewson, 2016) and droplet digital PCR (Martinez-Hernandez et al., 2019) all indicated that lytic pelagiphages were more frequently observed within oceanic ecosystems. Whole genome classification delineated these integrase-carrying, putatively lysogenic, phage from obligately lytic pelagiphages (without integrase) (Buchholz et al., 2021a, b; Zhang et al., 2021). It is also notable that all of the integrase-carrying pelagipodophages also carry a T7-like PolA having a tyrosine-762 (*Escherichia coli* numbering) amino acid residue hypothesized to be predictive of a phage lytic life cycle in bacteriophages that carry this gene (Keown et al., 2022; Nasko et al., 2018; Schmidt et al., 2014). These enigmatic features have stimulated hypotheses that prophage integration occurs within only small subpopulations of *Pelagibacter* or that these phage exhibit a lyso-lysis phenomena during lytic infection, as documented for coliphage lambda (Buchholz et al., 2021a, b; Zhao et al., 2019a). Nevertheless, molecular genetic evidence from cultivation studies and analysis of metagenomic sequence data indicates that integrase is active, and that pelagiphages integrate into SAR11 genomes (Zhao et al., 2019a). The fact that integrase-carrying pelagiphages multiply exponentially yet never seem to crash *Pelagibacter* cultures (Morris et al., 2020; Zhao et al., 2013) indicates that this virus–host relationship does not conform to either a strictly lytic or lysogenic life cycle.

Greater understanding of pelagiphage infection dynamics is critical in appreciating how virus–host interactions impact carbon flow through the ocean's most abundant bacterioplankton group, SAR11. If physical–chemical conditions are found to trigger switching between lytic and temperate phage propagation (Fig. 3), then environmental conditions could influence the strength of viral shunt or shuttle processes from abundant SAR11 populations in oceanic ecosystems under the influence of climate change. Most recent studies of pelagiphage isolates have focused on genomic characterization rather than culture-based characterization of infection dynamics such as burst size or latent period, particularly under different conditions of environmental parameters such as temperature, pH, or nutrient availability (Du et al., 2021; Zhang et al., 2021). Others have investigated genome content of metagenome-assembled phage genomes without isolation (Wittmers et al., 2022). Cultures of a lysogenic host, *Pelagibacter* sp. strain NP1, showed high host abundance even as the abundance of spontaneously produced viruses increased, but virus:host ratios grew to 0.84:1 or 15:1 under carbon-replete and carbon-deplete conditions, respectively (Morris et al., 2020). These data support that nutrient limitation could have

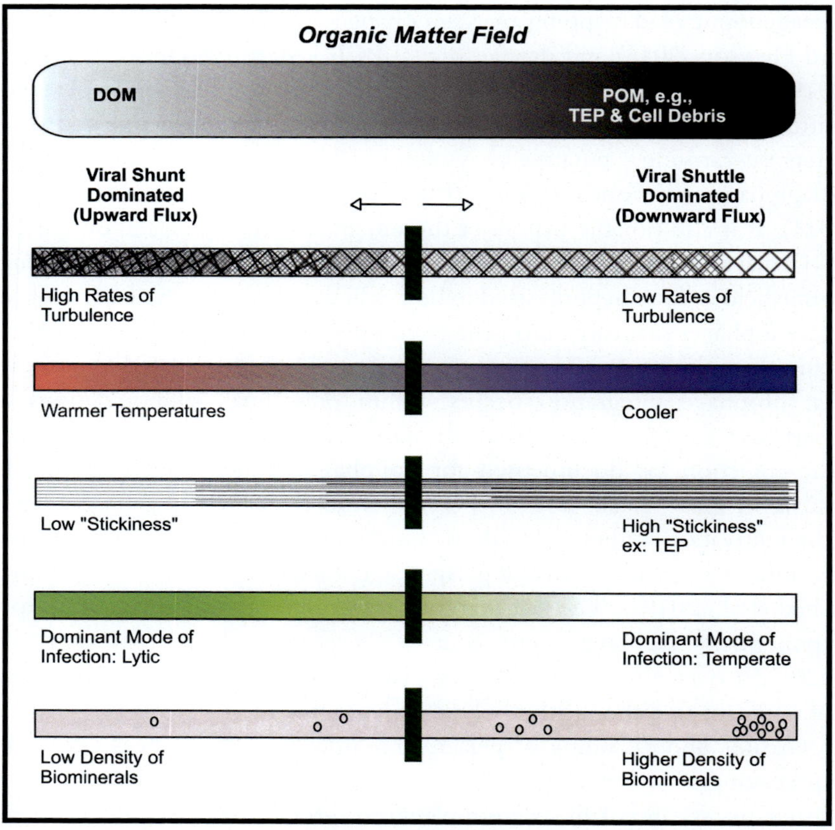

Fig. 3 Physical, biological, and chemical conditions toggle the switch between lytic and temperate phage propagation. Physical, biological, and chemical factors likely all influence whether lysis-mediated contributions to organic matter (OM) pools end up as dissolved OM (DOM) recycled by the viral shunt (upward flux) or as particulate OM (POM), e.g., transparent exopolymeric particles (TEP) or cellular debris, that are ultimately transported to the deep ocean via the viral shuttle (downward flux) for long-term carbon storage. DOM–POM interactions are likely influenced by both abiotic (e.g., temperature, turbulence) and biotic conditions (i.e. host physiology and virus–host interaction dynamics). Sliders at the center of each gradient indicate that each factor could push the system toward shunt (slide to the left) or shuttle (slide to the right). While factors are portrayed as independent gradients, they may either covary or interact synergistically to promote or suppress one carbon cycling pathway over the other.

consequential impacts on virus–host dynamics. Virus abundance was also measured in environmental samples using SYBR Green I chip-based digital PCR (SYBR dPCR) to enumerate particular, previously uncultured pelagiphage populations, with reported potential carbon release due to viral infection (McMullen et al., 2019). The evidence of differential host growth and infection under nutrient limitation, and likely virus shunt–shuttle environmental toggles, indicate a need to fill in the many gaps in understanding pelagiphage infection and its implications for nutrient cycling in our changing climate.

3.1.2 Roseophages

Roseobacter represent one of the major clades of marine bacteria, often comprising 15–25% of MLD bacterioplankton communities from coastal to open oceans (Wagner-Döbler and Biebl, 2006; reviewed in Buchan et al., 2005). Physiologically and morphologically diverse, these Proteobacteria are important players in carbon and sulfur cycling through a variety of adaptations as free-living, particle-associated or commensal cells capable of aerobic anoxygenic photosynthesis, CO oxidation, DMS production, sulfite oxidation, and the acquisition of energy for growth through both phototrophy and heterotrophy (i.e., mixotrophy) (Tang et al., 2010, 2016). However, few of these characteristics represent the entire clade, and instead subclusters occupy different marine ecological niches in coastal to pelagic oceans, tropical to polar ocean regions, and surface to seafloor and sediments. Roseobacter are found in algal blooms, microbial mats, and commensal relationships with marine phytoplankton, invertebrates, and vertebrates (Buchan et al., 2005; Geng and Belas, 2010; Jasti et al., 2005; Pohlner et al., 2017; Raina et al., 2009; Zhang et al., 2016, 2020). Predominantly marine, Roseobacter isolates and members of oceanic bacterioplankton communities have demonstrated a salt requirement or tolerance (Jonkers and Abed, 2003; Labrenz et al., 1998; Lau et al., 2004). Isolates can be representative of strains known to represent up to 20% of a given environment's natural bacterial community (Brinkmeyer et al., 2003; Fuhrman et al., 1994; Pinhassi et al., 1997; referenced in Buchan et al., 2005), but are poorly representative of natural diversity from other environments (Eilers et al., 2001; Selje et al., 2004; referenced in Buchan et al., 2005).

Given the environmental significance of their host, it is surprising that relatively few roseophage have been isolated with the first report occurring over 20 years ago (Rohwer et al., 2000). Isolated from multiple host clades, the majority of the >50 roseophage isolates (Bischoff et al., 2019; Huang et al., 2021; Zhan and Chen, 2019a; Zhang et al., 2019a) are dsDNA podoviruses (short tail) or siphoviruses (long non-contractile tail) with genome sizes ranging from 25 to 146 kb (Zhan and Chen, 2019a), though two ~4 kb ssDNA roseophage have been identified (Zhan and Chen, 2019b). In general, roseophage are thought to have a narrow host range as many isolates are only able to infect their original isolation host

(Chan et al., 2014; Yang et al., 2017; Zhang and Jiao, 2009), though there are several which are able to infect multiple host isolates (Li et al., 2016b; Zhao et al., 2009). Latent periods range from 1 to 6 h, with variable burst sizes from 10 to 1500 viruses (Cai et al., 2019; Li et al., 2016b; Rihtman et al., 2021; Zhao et al., 2009).

Roseophage are presumed to impact their hosts' contributions to carbon, sulfur, and phosphorus cycling, though little data is available on the comparison of phage infection dynamics under varying nutrient conditions. Instead, the potential for roseophage impact on nutrient cycling is primarily based on genome content and potential function. Whole genome comparative analysis of 50 roseophage generated 32 operational taxonomic units (OTUs) based on 97% nucleotide similarity (Huang et al., 2021). Auxiliary metabolic genes (AMGs), elsewhere referred to as host-derived genes, were identical within each OTU, and further divided into low frequency (45% of AMGs were present in a single genome OTU) and high frequency groups. High frequency AMGs include six genes related to nucleotide biosynthesis (RNR was present in 67% of genome OTUs) and a single gene, *phoH*, involved in phosphate intake. The first sequenced roseophage genome, Roseophage SIO1, was the first phage genome reported to carry *phoH*- or *thyl*-like genes (Rohwer et al., 2000). Low frequency genes were involved in multiple metabolic pathways including protein and receptor metabolism, cell signal transduction, and phage competition. Integrases or lytic repressors were identified in ~75% of lytic roseophage isolate genomes overall and in almost all roseosiphophage, suggesting the potential for integration into the host genome and a lysogenic lifestyle (Forcone et al., 2021; Zhan and Chen, 2019a).

There is also evidence of temperate roseophage based on induction with mitomycin C or UV exposure, and prophage and phage-like gene transfer agent regions in host genomes (Ankrah et al., 2014a; Chen et al., 2006; Forcone et al., 2021; Sonnenschein et al., 2017; Zhan et al., 2016; Zhao et al., 2010). Whole genome analysis of 79 host roseobacter bioinformatically identified 173 prophage regions (22 high-quality or complete) (Forcone et al., 2021). Though the most commonly observed gene was located in these prophage regions, all 22 instances were observed in low-quality regions and were considered to be gene transfer agents (GTA) based on synteny and homology to rcGTA, a GTA of *Rhodobacter capsulatus* strain 1003. Of the 50 medium- or high-quality prophage regions, only five carried an integrase. No replication or nucleotide metabolism genes, such as RNR, were observed in these prophage regions or in the genomes of two temperate roseophages (Ankrah et al., 2014a), supporting the hypothesis that RNR is absent from most temperate phage and is indicative of lytic replication (Harrison et al., 2019; Nasko et al., 2018). Read recruitment of *Tara* Oceans viromes to these 50 prophage regions suggests that abundance decreased with increasing latitude and temperature. Increasing temperatures due to global climate change could influence the prevalence of lysogenic infection and modulation between viral shunt and shuttle processes.

3.1.3 Cyanophages

Cyanobacteria encompass all prokaryotic taxa capable of oxygenic photosynthesis, making this group of microorganisms critical players in the global carbon cycle and especially the world's oceans. Cyanobacteria are divided into three morphological groups based on cellular aggregation characteristics and the presence of a unique differentiated cell type known as the heterocyst, a cell type dedicated to nitrogen fixation. These types are aggregated or solitary non-heterocystous filamentous, heterocystous filamentous, or single cells. Diverse physiological and ecological adaptations, including consortial and symbiotic associations, coloniality, nutrient sequestration and storage, buoyancy regulation, and nitrogen fixation, also reflect cyanobacteria adaptations to a wide range of environmental conditions shaped by nutrient depletion, turbulence, and suboptimal light and temperature (Mann and Clokie, 2012). As the only prokaryotes capable of photosynthesis, these diverse bacteria are restricted to the euphotic zone in the upper 100–200 m, and represent a significant portion of phytoplankton productivity and biomass. Small ($<3\,\mu m$) single-celled picocyanobacteria, most notably of the genera *Prochlorococcus* and *Synechococcus*, dominate vast expanses of the global oceans and often constitute $\sim$25% of oceanic photosynthetic production and biomass (Flombaum et al., 2013; Mann and Clokie, 2012). Oceanic cyanobacteria are also major players in the global nitrogen cycle, with outsized contributions from a free-living unicellular cyanobacteria (*Crocosphaera*), a group of uncultured, unicellular symbiotic cyanobacteria (UCYN-A), a filamentous non-heterocystous cyanobacteria (*Trichodesmium*), and groups of filamentous heterocystous symbionts (*Richelia* and *Calothrix*) (Zehr, 2011; Zehr and Capone, 2020).

Given the global abundance and significance of their hosts, cyanophages represent one of Earth's most abundant biological entities. Cyanophage infections shape the composition and availability of DOM as current estimates indicate that 5–40% of cyanobacteria are infected and lysed every day (Fuhrman, 1999; Proctor and Fuhrman, 1990). Viral infection impacts the chemical profile of available DOM pools, not only due to variations in DOM chemical composition produced by the lysis of different genera (Becker et al., 2014), but also through the release of DOM compounds different than those produced via exudation (Xiao et al., 2021) or mechanical cell lysis (Ma et al., 2018). For example, DOM produced by viral-mediated *Synechococcus* lysis contained higher molecular weight nitrogen-containing compounds when compared with mechanical cell lysis (Ma et al., 2018), and the virus-mediated release of intracellular iron occurs at a greater rate and with greater bioavailability compared to iron released from cyanobacteria without phage (Poorvin et al., 2004). Observations within oxygen deficient zones at the base of the euphotic zone indicate high cyanophage: cyanobacteria ratios and presumably high levels of cyanophage contribution to DOM release from sinking particles containing cyanobacterial host cells (Fuchsman et al., 2019).

Reflecting the diversity of their hosts, cyanophage across and within all host morphological groups are diverse in terms of both their genomes (lengths of 30–252 kb, GC content, and gene content and organization) and phenomes (morphology; infection dynamics including burst size, latent period, and diel patterns; and impact on biogeochemical cycling). Isolated cyanophage carry dsDNA genomes, although a ssDNA prophage-like particle was induced in *Synechococcus* cultures in response to mitomycin C (McDaniel and Paul, 2006). Cyanophage are present as myo- (long contractile tail), podo- (short tail), or siphovirus (long non-contractile tail) morphologies. Additional cyanophage morphologies have been observed, including filamentous freshwater *Microcystis* and *Anabaena* phage (Deng and Hayes, 2008) and tailless freshwater *Planktothrix* (Gao et al., 2009) and *Pseudoanabaena* (Zhang et al., 2020) phage. Most cyanophage are in the myo and podo morphological groups, with genome organization resembling T4-like and T7-like bacteriophages, respectively. Cyanosiphoviruses are found globally but in far lower densities, and have more diverse genomes (Huang et al., 2012).

Most cultivated phages are obligately lytic, and cyanophage are no exception. LPP-1, named for the *Lyngbya*, *Plectonema*, and *Phormidium* genera it infects, was the first cyanophage isolated, in 1963 (Safferman and Morris, 1963, 1964). Extensive collections of lytic cyanophage have been isolated infecting multiple host genera including *Anabaena* (Hu et al., 1981; Wu et al., 2009), *Nostoc* (Chénard et al., 2016; Hu et al., 1981),

Procholorococcus (Sullivan et al., 2003), and *Synechococcus* (Chen and Lu, 2002; McDaniel and Paul, 2006; Stoddard et al., 2007), and often from the specific well characterized *Synechococcus* isolate, WH7803 (Kana and Gilbert, 1987; Kana et al., 1988). Temperate cyanophage have been isolated infecting freshwater *Anabaena* (Franche, 1987; Marei et al., 2013), *Anacystis* (Lee et al., 2006), and *Synechococcus* (Chu et al., 2011; Dillon and Parry, 2008); and marine *Synechococcus*. Temperate cyanophage have been induced from cultures and natural populations of multiple genera such as *Trichodesmium* (Hewson et al., 2004; Ohki, 1999) and *Synechococcus* (including a ssDNA, tailless inducible particle from a *Synechococcus* isolate) (McDaniel et al., 2002; McDaniel and Paul, 2006; Ortmann et al., 2002). However, no intact prophage regions have been identified in cyanobacteria genomes, despite genomic remnants of phage integration into genomes of multiple genera including those for *Synechococcus*, *Prochlorococcus*, *Nostoc*, and *Anabaena* (Coleman et al., 2006; Flores-Uribe et al., 2019; Jungblut et al., 2021; Malmstrom et al., 2013; Qiu et al., 2019; Shitrit et al., 2022).

Filamentous cyanobacteria are critical players in biogeochemical cycling in many aquatic ecosystems. Phage isolates infecting these hosts provided the first cultivated model systems, yet few isolates have sequenced genomes and well characterized infection dynamics that permit the generation and testing of genome to phenome hypotheses with subsequent application to metagenomic datasets (Chevallereau et al., 2022). The outsized importance of cyanobacteria in the global carbon cycle makes understanding the effects of cyanophages on these microbial hosts critically important. Some filamentous cyanobacteria are known to form blooms, potentially releasing toxins with significant adverse health effects. Infection of *Nodularia spumigena* with cyanophage isolate 68v162–1 led to higher cellular concentrations of the hepatotoxin nodularin (Šulčius et al., 2018). These blooms are expected to be more common and intense with warmer temperatures due to climate change (Gobler, 2020; O'Neil et al., 2012; Paerl and Huisman, 2009).

Among the most astonishing findings characterizing cyanophage–host dynamics was the discovery that obligately lytic cyanophages carry a suite of host-derived metabolic genes that, when expressed in the infected cell (virocell; Forterre, 2011), serve in maintaining or enhancing key aspects of host metabolism. The first of these discoveries was a report that *Synechococcus*-infecting phage S-PM2 carries the *psbA* gene, encoding the photosystem II D1 protein (Mann et al., 2003). A flurry of other reports followed demonstrating that cyanophage photosystem genes were likely acquired from cyanobacteria (Sullivan et al., 2006) and are expressed during infection (Lindell et al., 2005), and that some *Prochlorococcus*-infecting cyanophages carried multiple photosystem genes (Sullivan et al., 2006). Intriguingly, those phages having a greater number of genes, beyond *psbA*, tended to also have broader host ranges (Sullivan et al., 2006). It is clear from virioplankton community studies that the phenomena of carrying photosystem genes is widespread among cyanophages having highly divergent genomic and phenomic characteristics (Sandaa et al., 2008). That cyanophages specifically carry photosystem II genes was rational as UV damage causes rapid turnover of these proteins. By encoding its own *psbA*, the infecting cyanophage maintains photosystem function and sustains photoautotrophy necessary for phage replication. The influence of cyanophage on photosynthesis can also extend to control of cellular metabolic pathways responsible for synthesizing key light-harvesting proteins (Gasper et al., 2017; Puxty et al., 2015; Shan et al., 2008).

Cyanophage, however, have acquired many other host-derived genes that presumably increase fitness during infection, enabling these phages to respond not only to the extracellular environment and particular niches inhabited by their hosts but to intracellular nutrient and substrate availability. Cyanophage at depth and lower light intensity carry genes encoding proteins within the purine synthesis pathway (*purN*, *purM*, *purC*, and *purS*) which may sustain purine production for nucleotide synthesis, energy storage and transfer, signaling, and cofactor production (Sullivan et al., 2005; Zhao et al., 2013). These genes promote continued host metabolism for the longer latent periods observed at low light levels rather than at the surface. Several genes (e.g., the gene encoding CP12, as well as genes *talC*, *zwf*, and *gnd*) direct carbon flux toward the pentose phosphate pathway for NADPH production and ultimately dNTP biosynthesis (Thompson et al., 2011). Phosphate uptake and regulation is impacted by viral genes encoding pstS, phoA, and phoH, members of the pho regulon (Rong et al., 2022). Cyanophage carry several proteins that seem to be confined to phages infecting cyanobacteria, including DNA polymerase gamma (Chan et al., 2011) and the Cyano M (cyanomyovirus) and Cyano SP (cyanosipho- and cyanopodovirus) clades of RNR (Harrison et al., 2019). The key lesson learned from the numerous studies reporting that cyanophages encode genes capable of sustaining or redirecting host central metabolism during infection is that cyanophages can substantially alter the spectrum of metabolites within the cell (Gao et al., 2016). Subsequent lysis alters the DOM composition released from cyanobacterial populations compared with that released from uninfected cells (Zheng et al., 2021). Alterations in DOM composition with viral-mediated lysis are critical for understanding carbon fate (shunt or shuttle) in the upper ocean (Fig. 1). It is the interplay of prevailing physical and chemical conditions around DOM release from cyanophage lysis that will shape whether this carbon recycles and remains within surface waters that readily exchange carbon with the atmosphere or aggregates and sinks below the mixed layer.

Cyanobacteria are expected to increase in abundance under predicted climate conditions such as increased temperature, stratification, and nutrient availability (reviewed in Carey et al., 2012). The ecophysiological adaptations of cyanobacteria across and within genera, including photosynthesis at multiple intensities and wavelengths, proliferation and bloom formation at warmer temperatures, buoyancy regulation, phosphorus storage, and nitrogen fixation, may allow them to dominate in many aquatic environments. Increases in cyanophage abundance have been observed with elevated seasonal temperatures (Clasen et al., 2013; Mankiewicz-Boczek et al., 2016; Marston and Sallee, 2003; Millard and Mann, 2006), though the opposite has been observed with particular clades (Maidanik et al., 2022). Many of the same environmental factors also impact cyanophage virulence (reviewed in Grasso et al., 2022). Cyanophage are stable under sustained increases in temperature often reflective of their hosts' temperature tolerances. While infectivity was observed to decrease with increased temperature (Garza and Suttle, 1998), cyanophage remained infective up to 45 °C (Safferman and Morris, 1964), with some thermotolerant strains infective after incubation at 70 °C for 60 min (Franche, 1987). Increased temperature is also found to impact infection dynamics, shortening the latent period and increasing burst size (Steenhauer et al., 2016; Yadav and Ahn, 2021). Induction of temperate cyanophage was observed with increasing temperatures (Chu et al., 2011; McDaniel et al., 2002; Rimon and Oppenheim, 1975), shifting phage impacts on nutrient cycling more toward viral shuttle than shunt.

Ocean stratification was correlated with cyanophage abundance and phylotype in natural populations, with minimal abundance at stable stratification (Maidanik et al., 2022). Stratification is associated with cyanobacteria community composition, with physiologically distinct *Prochlorococcus* and *Synechococcus* ecotypes observed in low light, high nutrient deep euphotic zone conditions compared to high light, low nutrient surface conditions

(Ahlgren and Rocap, 2006; Campbell and Vaulot, 1993; Partensky et al., 1996). Cyanophage abundance reflects this host stratification, with peak abundances occurring at the maximum abundance of *Prochlorococcus* ecotypes (Fuchsman et al., 2021) adapted to low light (near the base of the euphotic zone) and high light (surface waters). Cyanophage phylotype also correlated with stratification, with cyanophages carrying multiple photosystem genes dominant at the primary chlorophyll maximum depth and those carrying multiple nucleotide biosynthesis genes increasing with depth (Fuchsman et al., 2021). Although cyanophage diversity appears to positively correlate with stratification, dissecting the impacts of light, temperature, nutrient availability, and host abundance on this trend is difficult (Maidanik et al., 2022; Wilson et al., 2000).

Cyanophage abundance and infection dynamics are also impacted by nutrient availability. Phosphate depletion in cyanophage–host incubations increased latent period, decreased burst size, and decreased cell lysis, suggesting a switch to a more temperate life cycle under phosphate depletion (Cheng et al., 2019; Shang et al., 2016; Wilson et al., 1996), although conflicting impacts on viral adsorption were also observed. Phosphate addition in a mesocosm study resulted in an increase in viral abundance (Wilson et al., 1998), potentially reflecting an induction of temperate cyanophage. The observation of decreasing cyanophage titers along a transect from coastal to oligotrophic waters (Sullivan et al., 2003) also supports the hypothesis of a connection between phosphate availability and infection dynamics. Few studies have discussed the effect of nitrogen availability on cyanophage infection dynamics, though multiple studies have evaluated the impacts of cyanophage infections on the fate of nitrogen. Cyanophage infection has sustained nitrogen uptake and fixation, and impacted nitrogen flow through incorporation of extracellular nitrogen into new phage particles or the release of nitrogenous compounds with host lysis (Kuznecova et al., 2020; Pasulka et al., 2018; Waldbauer et al., 2019). More thorough experimentalist studies of cyanophage infection dynamics and life history will help unveil the potential carbon cycle impacts of cyanophage–host interactions in the changing ocean. In particular, improved understanding of the critical roles of key available nutrients, such as iron, phosphorus, and nitrogen (the abundance of which will certainly change alongside the broader physical and chemical changes in the ocean) will shape predictions of viral-mediated carbon release from oceanic cyanobacteria.

3.2 Viral interactions in phytoplankton groups critical in the oceanic carbon cycle

3.2.1 *Viruses of haptophytes—Coccolithophores and* Phaeocystis

The haptophytes, or prymnesiophytes, consist of ~500 living species (within 50 genera) of primarily marine unicellular, photosynthetic, globally distributed algae, some of which synthesize biominerals which are incorporated into the geological sediment record. Perhaps the best-known haptophytes are coccolithophores, which enhance transport of particulate organic carbon (POC) from the upper ocean to the seafloor through calcification, a light-dependent process that produces particulate inorganic carbon (PIC) in the form of calcium carbonate ($CaCO_3$) plates known as coccoliths (Paasche, 2001; Rost and Riebesell, 2004). Globally, coccolithophores account for at least half of the annual 80–120 Tmol of PIC produced in the pelagic ocean (Balch et al., 2007; Berelson et al., 2007; Broecker and Clark, 2009; Westbroek et al., 1993), with ~50% of this calcite reaching the sea floor (Broecker and Clark, 2009). Given that calcite is denser and experiences less water column dissolution than other ballast biominerals like biogenic silica, coccolith-associated calcite is responsible for up to 83% of the carbon flux to depth globally (Klaas and Archer, 2002). Thus, fluctuations in PIC and/or POC production by coccolithophores under different environmental conditions can impact associated $CaCO_3$:POC ratios (or "rain ratios") with important implications for the efficiency of the biological pump (Armstrong et al., 2001; Klaas and Archer, 2002; Ridgwell et al., 2009).

Arguably, the most direct evidence for stimulation of carbon export by virus infection comes from the globally distributed, cosmopolitan coccolithophore, *E. huxleyi*. *E. huxleyi* forms massive blooms spanning hundreds of square kilometers across diverse oceanic regimes excluding the polar oceans (Tyrrell and Merico, 2004), but that is likely to change with climate alterations. These blooms, which develop and then fade over ~10 days, are detected by Earth-observing satellites and are routinely terminated by lytic dsDNA-containing coccolithoviruses called EhVs (Bratbak et al., 1993; Vardi et al., 2012). As members of the Phycodnaviridae, EhVs are giant microalgal viruses (~180 nm in diameter) with an extensive genetic capability (~407 kb genomes) for manipulating host metabolic pathways for their replication (Castberg et al., 2002; Schroeder et al., 2002; Van Etten et al., 2002; Wilson et al., 2005). A variety of lipid- and nucleic acid-based biomarkers have been developed using both sensitive and resistant strains of

E. huxleyi and various strains of EhVs (Bidle and Kwityn, 2012; Bidle et al., 2007; Fulton et al., 2014; Hunter et al., 2015; Kendrick et al., 2014; Nissimov et al., 2019a, b; Rose et al., 2014; Rosenwasser et al., 2014; Schieler et al., 2019; Sheyn et al., 2016, 2018; Vardi et al., 2009, 2012; Ziv et al., 2016) in both lab-based model systems for which reference genome sequences and transcriptomic responses for hosts and viruses exist (Allen et al., 2006a, b; Nissimov et al., 2013, 2016, 2017; Wilson et al., 2005) and for natural populations (Hunter et al., 2015; Knowles et al., 2020; Laber et al., 2018; Sheyn et al., 2018; Vardi et al., 2009, 2012; Vincent et al., 2021). This has unlocked unprecedented insight into EhV infection strategies, ecological connections, and biogeochemical outcomes (Bidle, 2015; Bidle and Vardi, 2011). Direct evidence for viral stimulation of both POC and PIC export comes from the North Atlantic Virus Infection of Coccolithophores Expedition (NA-VICE) for which the aforementioned diagnostic markers were used to diagnose mesoscale (~50–100 km) *E. huxleyi* blooms within early-, late- and post-infection stages (Laber et al., 2018). Active, early infection of *E. huxleyi* blooms greatly increased export fluxes of both PIC and POC from the surface mixed layer into the mesopelagic, thereby increasing biological pump efficiency and "shuttling" carbon to depth aided by the dense, calcite biomineral. Carbon export fluxes down to 300 m were measured both using Particle Interceptor Traps (PITs) and optical profilers, equipped with backscatter, fluorescence, and oxygen sensors, through which spike signatures and associated carbon respiration were quantified into the mesopelagic (Laber et al., 2018). Preferential export and sinking of infected cells from the overlying surface waters was confirmed by the enrichment of infection-specific lipids (viral glycosphingolipids and betaine-like lipids) ratios in sedimenting particles, along with gene expression profiling (Sheyn et al., 2018). Active infection in sinking aggregates has since been corroborated using high-throughput single-molecule messenger RNA *in situ* hybridization (smFISH) in coastal populations (Vincent et al., 2021). Taken together, they provide "smoking gun" evidence for virus-mediated carbon export (Laber et al., 2018).

Carbon export is facilitated by virus-induced, cellular production of TEP, a sticky acidic polysaccharide with a density of ~0.8 g cm^{-3}, produced in response to viral infection (Nissimov et al., 2018). Under the right micro-scale physical conditions where particle encounters are accentuated (Box 1), the overproduction of sticky TEP by infected cells aggregates cells, cellular debris, and dense coccolith biominerals into larger aggregates with high sinking transport rates into the mesopelagic. Large-scale infection-induced coccolith shedding (Johns et al., 2019), as well as recent surprising findings that coccolith biominerals interactively adsorb both cells and viruses (C.T. Johns, Personal communication, 2022), enhance infections and entrain large pools of ballast biominerals into aggregating particles, further facilitating export to depth with important implications for carbon sequestration over long geologic timescales.

Despite the availability of a unique suite of biomolecular tools for diagnosing coccolithophore infection at sea, predicting the outcome of infection events in terms of attenuation and rapid respiration of carbon in the surface ocean (virus shunt) or aggregation and export of carbon into the mesopelagic (virus shuttle) remains a challenge (Fig. 1). Predicting these outcomes is critical, because they determine whether carbon remains active or is sequestered for hundreds to thousands of years depending on sinking depths reached. A few mechanisms have surfaced in the *E. huxleyi* system that can toggle these respective outcomes of bloom demise. These mechanisms serve to critically inform our thinking and predictive understanding in our virus–host systems.

The first mechanism is the relative ability of EhV strains to induce sticky exopolymer production (e.g., TEP) and stimulate particle formation. EhV strains can have different infection dynamics in culture, based on their regulation of lipid metabolism and the nature and activity of their encoded serine palmitoyl transferase (SPT) enzymes, which impact host cell responses to viral infection, TEP production, and aggregation (Nissimov et al., 2018, 2019a, b). For example, fast-infecting, more virulent viruses generate more TEP and their characteristic SPT gene signatures are more enriched in sinking particles found at depth (~150 m). Conversely, slow-infecting, less virulent EhVs are more widely distributed in surface populations and less connected with exporting particles (Nissimov et al., 2019a, b), despite a wider host infection range. It is critical to note that in these examples the differing life history strategies of the infecting EhV changed the fate of carbon existing within *E. huxleyi* biomass.

Secondly, the fate of carbon during infection is fundamentally impacted by the mode of infection itself. Due to a lack of diagnostic infection markers, limited understanding of infection dynamics has fundamentally hampered our ability to discern how virioplankton populations actually impact the fate of carbon in natural systems. Another confounding factor has been the traditional emphasis on studying virus–host systems in lab cultures that exceed virus and host concentrations by orders of magnitude over those observed in natural systems. The *E. huxleyi*–EhV system, a traditional paragon of virulence, was recently shown instead to be characterized by fundamentally different, temperate infection dynamics (Knowles et al., 2020). Rather than showing density-dependence, EhVs exhibit extended asymptomatic infection during bloom formation (virus–host "*Détente*"), only killing their hosts when they become stressed upon resource limitation (a viral-mediated "*Coup de grâce*"), suggesting that viral propagation and bloom termination are physiology-dependent rather than density-dependent processes. Corroborating evidence for the chronic release of particles from *E. huxleyi* cells in the absence of lytic infection was also recently demonstrated in natural, coastal populations (Vincent et al., 2021). This *Coup de grâce* model challenges long-held tenets in oceanography and viral "rules for infection strategy" with broad implications. Temperate infection in *E. huxleyi* may explain how EhV infection is sustained despite the encounter rate gap (K. Bondoc, Personal communication, 2022) between free viruses and *E. huxleyi* cells existing in the dilute ocean conditions. Temperance could aid in spreading EhV across micro- (microns) to meso-scales (kilometers) to collapse blooms the size of small countries, particularly under environmental conditions that present barriers to virulent transmission like low virus–host encounters, rapid viral decay, and absorptive particles. These barriers apply universally to viruses infecting oceanic microbial hosts. Thus some measure of "temperate" phenotypes outside of the classic models identified for well-known experimental virus–host systems, as possibly seen in pelagiphages carrying integrase genes (Zhao et al., 2019b), may be essential to the propagation of many oceanic viral populations. Beyond EhV dynamics, it is likely that other traditionally strictly "virulent" models actually show environmentally and physiologically induced temperate behavior. Transitioning from density- to physiologically-driven viral controls of bloom formation and decline is also a major shift in how we view and model bloom growth and death processes. Such a relationship resembles conditional symbiosis (Roossinck and Bazán, 2017), which can enhance host proliferation and reduce predation, with transition to lethal lytic stages happening when host cells become physiologically stressed and the symbiosis breaks down. It highlights the need for conceptual and experimental emphasis on coupled virocell interactions at the microscale driving bloom formation and propagation across mesoscales based on altered host physiology rather than virus–host densities.

Lastly, the mechanisms and degree to which virus infection couples with grazing activity can significantly shape carbon cycling outcomes. TEP plays an outsized role in this context. Because TEP facilitates aggregation and stimulates sinking of infected particle aggregates, it can also serve to couple infection and grazing by shifting the availability of infected cells towards larger, macrozooplankton grazers which produce fecal pellets with very high sinking speeds (hundreds of $\mathrm{m\,d^{-1}}$). This process further shuttles virus-infected cells to depth. Direct evidence for this phenomenon in natural populations came by examining the gut contents of copepods for the presence of infected *E. huxleyi* cells. EhV major capsid protein genes were detected in nearly all collected copepods (Frada et al., 2014). Subsequent production of fecal pellets further enhances export of infected material, essentially "lubricating" the biological pump for greater efficiency in carbon export. Observations of virus–zooplankton coupling suggested that such processes may also serve as possible transmission vectors for virus infection in the upper ocean as macrozooplankton (e.g., copepods) swim and vertically migrate through the water column (Frada et al., 2014).

Another cosmopolitan, bloom-forming haptophyte genus that is a key contributor to primary productivity and biogeochemical cycling in high latitude and polar environments is *Phaeocystis*. This unicellular algal genus displays a unique ability for cycling between single, flagellated cells and colonial, non-flagellated cells. This physiology is particularly relevant to carbon cycling and the coupling of virus infection and grazing as relevant carbon loss factors. In their colony form, prominent *Phaeocystis* species (e.g., *P. pouchetii* and *P. globosa*) routinely dominate phytoplankton communities, and field studies indicate that virus infection is a dynamic component involved in the decline of *Phaeocystis* blooms (Brussaard et al., 2007).

Viruses that infect species of *Phaeocystis* have been isolated in natural blooms (Baudoux and Brussaard, 2005; Brussaard et al., 2004; Jacobsen et al., 1996) and, like EhVs, belong to the Phycodnaviridae family of nucleocytoplasmic large DNA-containing viruses (NCLDVs) (Van Etten et al., 2002). During infection *Phaeocystis* viruses induce morphological, physiological, and viability changes in their host populations (Bratbak et al., 1998b; Brussaard et al., 2001; Jacobsen et al., 1996). Mesocosm and field studies on natural populations provided evidence that viruses could be important mortality agents for *P. globosa* and *P. pouchetii* as the concentrations of *Phaeocystis* viruses exceeded respective host concentrations by 30- to 100-fold during bloom maxima (Brussaard et al., 2004, 2005a; Larsen et al., 2001). This phenomena is heavily impacted by *Phaeocystis* life cycle stage, with the single-cell stage of *P. globosa* supporting 30% increases in *P. globosa* virus (PgV) concentrations, as compared to the colonial form, resulting in prevention of a *P. globosa* bloom.

Observations of *P. globosa* bloom dynamics has implicated cell lysis being responsible for loss rates of up to 30% d^{-1} (Brussaard et al., 1995), with viruses being the most likely cause (Baudoux and Brussaard, 2005; Brussaard, 2004; Brussaard et al., 2005b; Larsen et al., 2001). In cultivation studies, viral lysis rates of *Phaeocystis* cultures have been observed as high as 80% d^{-1}

(Brussaard et al., 1999). Unfortunately, methods for specific and accurate determination of viral-mediated mortality of *Phaeocystis* (and other key phytoplankton genera) in natural waters are still lacking. Adaptations of the classical grazer dilution method (Baudoux et al., 2006), originally designed to determine microzooplankton grazing rates (Evans et al., 2003; Landry and Hassett, 1982), have shown limited success with large error in viral lysis estimates (see Section 4.2 for more information). Other mortality estimates have relied on virus production and loss rates, combined with assumed burst sizes and measurements of total cell lysis (Baudoux et al., 2006). Observations of *P. globosa* blooms have showed microzooplankton grazing dominating cell loss during bloom development, while viral lysis became increasingly important at later stages, with rates comparable to grazing (as high as 35% d^{-1}) and accounting for over 50% of the total loss of single-cell morphotypes (Baudoux et al., 2006). Modeling also supports the dominance of viral lysis as the principal loss of *P. globosa* single cells (Brussaard et al., 2007). Colonies appear to be protected against grazing and viral infection, allowing for biomass accumulation and bloom development. Indeed, modeled carbon budgets showed *P. globosa* dominated primary production only when colonies were present (68% of total). Moreover, daily carbon flux from viral lysis was 10-fold higher with only single cells present as compared with conditions that included colonies (115 vs. 11 $\mu g\,C\,L^{-1}$). Viral lysis respectively accounted for 2% and 53% of daily *P. globosa* primary production with and without colonies present.

Energy flow and biogeochemical cycling within pelagic and benthic ecosystems where *Phaeocystis* is an important phytoplankton group is highly influenced by the fate of *Phaeocystis* primary production (Schoemann et al., 2005). Extensive viral lysis provides a source of DOM and regenerated inorganic nutrients in the euphotic zone (i.e., the viral shunt) (Brussaard et al., 1996; Gobler et al., 1997; Ruardij et al., 2005; Wilhelm and Suttle, 1999). In *P. pouchetii* cultures, viral lysis rapidly (within 3 days) drove algal biomass into DOC (Bratbak et al., 1998a). Observations of *Phaeocystis* blooms showed that viral lysis stimulated bacterial secondary production (Brussaard et al., 2005a), with carbon release from *Phaeocystis* cell lysis accounting for bacterial carbon demand. Observed shifts in responding bacterial community composition resulted in differential DOC fate, such that sudden lysate bursts of readily degradable and organic nutrient-rich DOC favored opportunistic bacterioplankton populations (Brussaard et al., 2005b). Similar observations of efficient trophic coupling have been made in model ecosystems with *P. pouchetii*, *P. pouchetii* virus (PpV), bacteria, and heterotrophic nanoflagellates (HNF). Infection of *P. pouchetii* by PpV had strong positive effects on the concentrations of bacteria and HNF, with the mass balances of carbon, nitrogen, and phosphorus implicating efficient heterotrophic biomass transfer upon viral lysis (Haaber and Middelboe, 2009).

An intriguing finding related to carbon flow and possible coupling of viral lysis to grazing is that viral infection of *P. globosa* impedes the formation of carbon-enriched chitinous star-like structures, as visualized in single cells by high resolution nanometer-scale, secondary-ion mass spectrometry (nanoSIMS) and atomic force microscopy (AFM) (Sheik et al., 2013). Uninfected cells transfer up to 44.5 mmol C L^{-1} (36%) of cellular biomass in the form of these structures, implicating the star-like structures as having an important role in cell survival. Viral infection impedes the release of these structures and facilitates the formation of aggregated flocs after cell lysis which make infected cells more susceptible to grazing and possibly accentuates the virus shuttle.

3.2.2 Viruses infecting unicellular chlorophytes—Micromonas *and* Ostreococcus

Chlorophytes consist of cosmopolitan, unicellular picoeukaryotic (<2 μm in diameter) taxa of the green plastid lineage (Lewis and McCourt, 2004) found in both coastal and oceanic systems (Foulon et al., 2008; Guillou et al., 2004). Although often less abundant than prokaryotic picophytoplankton (e.g., marine cyanobacteria *Prochlorococcus* and *Synechococcus*) and generally not forming massive mesoscale blooms (unlike the aforementioned haptophytes), these picoeukaryotes can account for more biomass per cell (6.5- to 14-fold more carbon) and can exhibit higher growth rates than the evolutionarily distant picophytoplankton. Because of this biomass difference, picoeukaryotes account for 76% of the net carbon production in oceanic ecosystems, while helping to facilitate carbon transfer to higher trophic levels through grazing (Worden et al., 2004). Understanding the turnover of picophytoplankton biomass through viral infection is important for understanding the global carbon cycle as picoeukaryotes account for such a large portion of carbon production (Cottrell and Suttle, 1995a). In some cases, viral lysis has been estimated to consume 9–25% of picoeukaryote standing stock each day (Evans et al., 2003). Species within two major chlorophyte genera, *Micromonas* and *Ostreococcus,* have been used as model organisms for exploring virus–host infection dynamics within marine chlorophytes, with most of the experimental work focused on *Micromonas* virus–host systems.

In fact, the first reported dsDNA virus infecting a marine eukaryotic phytoplankton infects a *Micromonas* host (Mayer and Taylor, 1979).

Studies have explored some of the factors that impact infection dynamics within chlorophytes, including light (Baudoux and Brussaard, 2008; Brown et al., 2007; Derelle et al., 2018; Zimmerman et al., 2019); UV-radiation (Eich et al., 2021); nutrients including nitrogen, phosphorus, and iron (Bachy et al., 2018; Maat and Brussaard, 2016; Maat et al., 2014; Slagter et al., 2016); as well as temperature (Demory et al., 2017). In *Micromonas* cultures adapted to different light regimes (25, 100, and 250 mmol photons $m^{-2}\,s^{-1}$), the impact on infection dynamics was relatively minimal, with all treatments undergoing cell lysis within 10–20 h post infection with similar burst sizes (285–360 viruses $cell^{-1}$) (Baudoux and Brussaard, 2008). Similar results were shown for *Ostreococcus*, however reduced burst sizes were observed at lower light levels (15 mmol photons $m^{-2}\,s^{-1}$) in two different virus strains (Zimmerman et al., 2019). Complete darkness inhibits viral replication (Baudoux and Brussaard, 2008), potentially from reduced viral adsorption to cells, which was also observed in the *E. huxleyi*–EhV system (Thamatrakoln et al., 2019). However, when viruses do infect, light impacts the timing of viral gene expression, with 96.8% of predicted viral genes transcribed at night in the dark (Derelle et al., 2018) albeit only observed in *Ostreococcus* thus far. UV exposure is also expected to play a role in infection dynamics, especially with intensified vertical stratification predicted due to climate change, and it has been shown that greater than 28 h of exposure to UV-AB wavelengths lowered viral production in a *Micromonas polaris* virus–host system (Eich et al., 2021).

Nutrient limitation (i.e., reduced nitrogen and phosphorus) has a mixed impact on the latent period of infection in a *Micromonas pusilla* virus–host system (Maat and Brussaard, 2016; Maat et al., 2014), with nitrogen-limited cells showing no change (Maat and Brussaard, 2016), and phosphorus-limited cells showing a longer latent period in some cases (Maat et al., 2014). However, under both nitrogen and phosphorus limitation, there was lower viral production yielding lower burst sizes (Maat and Brussaard, 2016), with some examples citing a lower percentage of lysed cells under phosphorus limitation (Bachy et al., 2018). Iron limitation also lowers burst sizes, as well as reduces the infectivity of viruses by ~30% (Slagter et al., 2016). In addition, temperature has an important impact on infection dynamics. When cells were grown below their predicted optimum temperature, infections took longer with reduced viral production, whereas cells grown above their optimum temperature did not undergo cell lysis

(Demory et al., 2017). Similar results were observed in the *E. huxleyi*–EhV system, where cells grown at higher temperatures (i.e., 21 °C instead of 18 °C) were not infected (Kendrick et al., 2014). Although these were reductionist experimental systems, they nevertheless illustrate that a few degrees of increased temperature (which is an increasingly common occurrence as the ocean changes along with the climate) can have a dramatic impact on virus–host interactions.

Ultimately, a better characterization of infection dynamics and viral life strategies under varying conditions will help us understand the role that chlorophyte infection plays in regulating the flow of carbon between the shunt (carbon retention) and shuttle (carbon export) (Fig. 1). Some viral mortality studies in *Micromonas* have suggested that viruses can coexist stably within *Micromonas* populations (Cottrell and Suttle, 1995b) instead of inducing rapid cell lysis (Zingone et al., 1999). These observations suggest a mode of viral infection that is more temperate than virulent in nature. This would support the observation mentioned above that the *E. huxleyi*–EhV system shows a more temperate infection strategy at environmental host concentrations instead of the virulent dynamics shown in lab culture experiments (Knowles et al., 2020). While it has been shown that infection increases the release of DOM, there is also evidence demonstrating that TEP production increased during viral infection, similar to what has been observed in the *E. huxleyi*–EhV system (Nissimov et al., 2018). This increased TEP production can promote particle aggregation and sinking (Lønborg et al., 2013), a key factor to increasing biological pump efficiency. This provides some lab-based evidence that infection could help stimulate carbon export; however, as TEP production was only measured in *Micromonas,* these observations will need to be expanded to other chlorophytes, like *Ostreococcus*, to determine if any genera-specific variability exists among chlorophytes. Field work recently demonstrated a potential link between chlorophyte infection and carbon export. One study examined the impact of viral community composition (using the *Tara* Oceans dataset; Guidi et al., 2016) on carbon export efficiency and found that viruses infecting chlorophytes were positively correlated with carbon export efficiency (Kaneko et al., 2021). Overall, these studies highlight the potential link between infection of these picoeukaryotic phytoplankton and carbon export. They also highlight the need for expanded lab-based studies using other genera and species of chlorophytes to better understand genera/species-specific responses during infection, as well as more field-based observations to improve understanding of virus–chlorophyte infection dynamics in natural systems. This could be achieved with the development of better diagnostic biomarkers like the lipid-based biomarkers in the *E. huxleyi*–EhV system (Laber et al., 2018; Vardi et al., 2009) for interrogating the infection state of chlorophyte populations. Expanded lab- and field-based observations will help in interpreting the role of chlorophyte viruses in regulating biological pump efficiency and their impacts on the carbon cycle.

3.2.3 Diatom viruses

Diatoms are among the most widely distributed and diverse eukaryotic phytoplankton in the global ocean (Malviya et al., 2016), contributing up to 40% of total marine productivity (Field et al., 1998). Despite their global dominance and importance in the global carbon cycle, it took 15 years after the discovery of high viral abundance in the ocean (Bergh et al., 1989) for the first diatom virus to be discovered (Nagasaki et al., 2004). Once considered immune to viral infection due to the physical protection of their silica-based cell walls, the discovery of diatom-infecting viruses revealed a unique and unclassified group of marine viruses. Unlike bacteriophages and viruses infecting eukaryotic phytoplankton groups, diatom viruses are unique with genomes comprised of either single-stranded ssRNA or ssDNA and virions among the smallest ever observed (~20–40 nm in diameter; Nagasaki et al., 2004; Tomaru et al., 2015). These characteristics contrast greatly with the large capsid sizes and equally large dsDNA genomes of the "giant" viruses infecting haptophytes. Diatoms disproportionately contribute to carbon export owing to their biomineralized, silica-ballasted cell wall. Likewise, viruses that infect and subsequently cause diatom mortality can disproportionately impact the global carbon cycle.

Around 30 distinct ssRNA or ssDNA diatom viruses have been discovered and isolated from both marine and freshwater environments (reviewed in Arsenieff et al., 2022). These viral genomes range from 5 to 9 kb in length and contain between two and four predicted genes, which encode for a replication enzyme (e.g., RNA-dependent RNA polymerase for the ssRNA viruses or replicase for the ssDNA viruses), a viral capsid structural protein, and one or two unknown proteins. All known diatom viruses exhibit lytic infection, causing host lysis and mortality within 2–10 days of infection. However, evidence for the production of viruses prior to host lysis (Kimura and Tomaru, 2015; Shirai et al., 2008; Tomaru et al., 2014) suggests that a chronic or lysogenic life cycle may potentially exist among diatom-infecting viruses. Given the limited studies on lysogeny in diatom viruses, we will focus here on lytic infection and its subsequent impacts on carbon cycling.

With burst sizes ranging from 10^1 to 10^5 viruses produced per host cell, diatom virus abundance could equal or exceed the abundance of all other algal viruses combined. However, unlike these other systems, diatom viruses do not contain auxiliary metabolic genes like cyanophages (Hurwitz and U'Ren, 2016; Thompson et al., 2011) and roseophages (Huang et al., 2021). Similarly, there is no evidence for coordinated host and viral gene expression as seen in the *E. huxleyi*–EhV system (Sheyn et al., 2018). Thus, the impact of diatom viruses on the fate of diatom organic matter and associated elements is likely dictated by changes in host metabolism/physiology and/or the dynamics of infection, both of which can tip the balance between viral lysis as a shunt or shuttle of diatom biomass.

Recent studies highlighting the dynamic nature of diatom virus–host interactions suggest environmental conditions and host physiology play a critical role in determining whether viruses act as shunts or shuttles. Nutrient availability is an important factor influencing the timing of viral-induced host lysis and mortality. It was found that silicon and iron limitation had distinct and opposing impacts on infection of both natural diatom communities and laboratory virus–host model systems (Kranzler et al., 2021, 2019). As obligate silicifyers, diatoms require silicon for cell wall, or frustule, synthesis. Diatoms respond to short-term silicon limitation by decreasing biogenic silica production in favor of maintaining maximum growth (Brzezinski et al., 1990; McNair et al., 2018), but after prolonged limitation, undergo cell cycle arrest and eventual physiological stress and mortality. Diatom viruses appear to capitalize on this weakened state accelerating the latent period and time to host lysis (Kranzler et al., 2019), suggesting that infection under silicon-limiting conditions would favor the viral shunt. In contrast, iron limitation of diatoms appears to slow infection, leading to a longer latent period and significantly delaying and reducing host mortality (Kranzler et al., 2021). Temperature also impacts diatom virus infection dynamics in *Chaetoceros tenuissimus* virus–host systems (Tomaru et al., 2014). Higher temperatures accelerated viral-induced mortality for a ssDNA virus infecting all *C. tenuissimus* strains but showed differential strain effects for an infecting ssRNA virus. Here again, a critical physical parameter, temperature, predicted to increase within the oceanic regions diatoms inhabit, can have strain-specific impacts on virus–host interactions and likely knock-on effects on carbon cycling from this important phytoplankton group.

Working counter to the processes that might facilitate the viral shunt are those that would favor the viral shuttle. Similar to *E. huxleyi*, viral infection of diatoms stimulates particle aggregation, although diatoms aggregation processes are mediated by the production of proteinaceous, Coomassie-stainable particles, rather than TEP (Yamada et al., 2018). An additional mechanism facilitating the viral shuttle are recent reports that viral infection induces spore formation in *Chaetoceros socialis* (Pelusi et al., 2021). Spores represent a resting stage of diatoms characterized by a heavily silicified cell wall that has been associated with mass export events in the North Atlantic (Rynearson et al., 2013). The induction of spore formation by viral infection appears to be a defensive strategy as "infected" spores failed to propagate or transmit infectious viruses upon germination. Nevertheless, this phenomena adds to a growing list of virus-induced physiological changes in host cells that may have significant impacts on oceanic carbon cycling.

3.2.4 Dinoflagellate viruses

Among major eukaryotic phytoplankton groups, dinoflagellates are unique being well-characterized as mixotrophs and contributing to the carbon cycle as both autotrophic primary producers and heterotrophic secondary consumers. Despite the ubiquity, diversity, and abundance of dinoflagellates in marine environments, *Heterocapsa cicrularisquama*, a dinoflagellate capable of forming harmful blooms, is the only host for which viruses have been identified. Two genomically distinct viruses have been identified to infect *H. cicrularisquama*—a dsDNA virus, HcDNAV (Tarutani et al., 2001), and a ssRNA virus, HcRNAV (Tomaru et al., 2004). Both viruses remain taxonomically unclassified, but both induce host lysis in culture and in natural communities (Takano et al., 2018; Tarutani et al., 2001; Tomaru et al., 2004). The appearance of thick-walled cysts in infected cultures and successful regrowth has led to the hypothesis that a virus-induced shift in life stage may represent a host defense strategy (Nagasaki et al., 2003; Tarutani et al., 2001; Tomaru et al., 2004) as has also been proposed in diatoms (Pelusi et al., 2021). Fast-sinking dinoflagellate cysts have been observed to dominate sinking material (Heiskanen, 1993), thus representing another link between viral infection and carbon export. Furthermore, a recent study estimates dinoflagellates may contribute up to 35% of sinking carbon in certain regions of the ocean (Juranek et al., 2020). With predictions that harmful algal blooms, particularly those dominated by dinoflagellates, may become more prominent in a future ocean (Brandenburg et al., 2019; Glibert et al., 2014; Gobler et al., 2017), more in-depth studies are critically needed on physical encounter rates of hosts and viruses (given dinoflagellates actively swim), as well as how viruses impact growth, life stage, and mortality of this severely understudied group of phytoplankton.

Each of the aforementioned virus–host players influence the bioavailability, location, biochemical form, and fate of light-derived carbon. However, virus–host dynamics occur within a complex landscape of physical, biological, and chemical contexts. These contexts can be divided into three main components: (1) physical processes, e.g., how particles move and interact (Box 1); (2) organic matter biochemistry from cell lysis, e.g., ballast, density of individual OM components, and DOM (labile) vs. POM (recalcitrant) organic matter; and (3) biological contexts of the community, e.g., community composition of both hosts and viruses, and viral life history (Fig. 1). Even under predictable and well-established climate conditions, these components can individually and interactively influence both the probability of virus–host interactions, and the probability of organic material from cell lysis contributing to either the viral shunt or the viral shuttle (Fig. 3 and Boxes 1 and 2). Under climate change regimes, the carbon cycling consequences of changes in physical, chemical, and biological context are even more elusive to track, characterize, and mimic experimentally. Therefore, detecting generalizable patterns in virus-mediated carbon cycling under climate change conditions requires that marine virology must integrate improved viral infection dynamics data into biogeochemical modeling.

4. Modern approaches to investigating virus–host dynamics in a changing climate

Given that life history strategies underpin infection dynamics, understanding how viral life history strategies alter carbon flow is critical to connecting genomic data to ecological forecasting. Leveraging 'omic approaches and global modeling approaches in marine virology will likely lead to improved forecasting of virus–host responses to climate change. For example, data mining well-established metagenomes libraries for replication marker genes experimentally proven to correlate with life history strategies could provide real-time evidence for fluctuations in favored life history strategies under different oceanic conditions (seasonality, bloom associations, temperature variation, storm or upwelling events, etc.). Integrating those predicted life history strategies with current climate models forecasts of changing climatic conditions may support predictions of virus shunt- or shuttle-dominance in future oceans. Here, we highlight the potential for expanded 'omics toolboxes and modeling approaches to shed light on population dynamics and function, forecast ecological changes in viral life history strategies, and consequently better forecast the role of viruses in mediating carbon flux and fate at a global level.

4.1 Using the meta-omics toolbox for understanding virioplankton carbon cycle dynamics

Viral life history strategy, the physiological changes occurring in virocells, and the unique character of OM released from infected and lysed cells all play an important role in shaping the flow of carbon through oceanic microbial communities. Measuring the details of these three phenomena—viral life history, virocell physiology, and OM release—seems an intractable problem given the vast diversity of oceanic virus–host systems. However, modern 'omics tools now provide unprecedented capacity for observing virus and host communities at the population scale. Today, using high-throughput DNA sequencing technologies, viral oceanographers can identify unknown viral populations (Bin Jang et al., 2019; Roux et al., 2019), the hosts they may infect (Ahlgren et al., 2017), each population's potential functional capabilities through the genes they carry, and even a population's gene activity. This suite of approaches is known as metagenomics when sequencing viral genomic material (DNA and RNA) and metatranscriptomics when sequencing host cell messenger RNA. With high-resolution mass spectral analyses, viral oceanographers can track real-time changes in the organic matter released from infected cells or from cell lysis. Collectively these approaches are referred to as metabolomics. Continuous technological improvements and reductions in analytical costs have steadily widened the application of metagenomics and metabolomics approaches in viral ecology, but there is still much work to be done in fully leveraging these emerging technologies for understanding how virus–host interactions shape oceanic carbon cycling.

Continuous technological advancement, the increasing availability of high-throughput DNA sequencing, and bioinformatics algorithms for sequence analysis have transformed 21st century virioplankton research. Each sequencing technology and experimental approach provides unique advantages in studying virus–host systems and virioplankton communities. DNA sequencing technologies can be divided into two broad groups based on sequence read length and accuracy, features that to-date have been analytical tradeoffs. Instruments with short read lengths (i.e., less than ~250 bp) often have the highest accuracy and provide the largest volume of data per analytical run. These instruments also have the lowest per bp cost. These features have made short-read technologies attractive for metagenomic analyses, where massive data volume provides deep sampling of even rare populations within the virioplankton. However, accuracy and affordability come at the expense of observing true biological sequences. Each individual short read typically provides too little of its parent gene sequence for reliable assessment of gene identity and function. In other words, the information content of each short read is insufficient for most purposes in studying the genetic potential of unknown viral populations (Wommack et al., 2008). However, large metagenome libraries of individual short sequence reads can provide useful ecological and biological information if the reads are "mapped" (matched by sequence homology) to longer sequences such as whole viral genomes or long contiguous genome fragments (contigs).

Thus, an essential step in fully leveraging the information content of a virome (viral metagenome) is assembly of the short read library into contigs. Bioinformatic algorithms for short-read sequence assembly have seen dramatic improvements in speed and computational efficiency and are freely available (Boisvert et al., 2012; Ji et al., 2017; Li et al., 2016a; Liang and Sakakibara, 2021; Namiki et al., 2011; Nurk et al., 2017; Peng et al., 2012). The assembled contigs, some even being complete viral genomes, are useful for defining unknown viral populations and their genetic potential (Roux et al., 2019). Standard best practices in defining the minimum information for an uncultivated viral genome (MIUViG) (Roux et al., 2019) aid all researchers in leveraging contigs assembled in prior studies for new investigations. Subsequent mapping of short-read sequence virome libraries against contigs and complete viral genomes provides ecological information on the biogeographic and/or temporal prevalence and frequency of viral populations. The key assumption behind this analytical approach is that the number of short reads mapping to a contig indicates the abundance of the viral population represented by the contig. For example, mapping of short-read virome sequence libraries to the genome sequences of pelagiphages demonstrated the broad biogeographic distribution and high frequency of these phages in the global ocean (Buchholz et al., 2021a, b; Zhang et al., 2021; Zhao et al., 2013). In an analogous way, short-read metatranscriptome libraries of mRNA isolated from host cell communities have been mapped to phage genomes and contigs for assessing the lytic activity of phage populations (Alonso-Sáez et al., 2018). However, the scientific utility of short-read sequencing has limits in that only a small percentage of virome contigs (~10% or less) are of sufficient length (>5–10 kb) for assessing the genetic content of unknown viral populations. Moreover, sequencing of single virus particles purified directly from water samples showed that microdiversity within viral populations may be a reason that short-read sequence assembly fails in providing greater yield of long virome contigs (Martinez-Hernandez et al., 2017).

Fortunately, alternative long-read sequencing technologies can overcome some short-read sequencing limitations. Long-read sequencing approaches yield more informative virome sequences that have been useful for examining virioplankton population microdiversity (Warwick-Dugdale et al., 2019) and gene diversity within functional genomic modules of unknown viral populations such as replication modules (Nasko et al., 2018). However, the combination of long and short reads from the same virioplankton DNA sample is essential as long-read technologies have error rates approaching 10%. The most powerful approaches have used accurate short reads for "correcting" sequencing errors in long reads, however, some have shown that correction can be done without short reads (Beaulaurier et al., 2020). Another limitation of long-read sequencing is the need for larger input quantities of virioplankton DNA for sequencing, which requires greater effort in water sample processing. While short-read sequencing is possible from even sub-nanogram amounts of input DNA, long-read sequencing typically cannot go below a microgram (without amplification). The greater DNA requirements are, in part, a consequence of the fact that for a given quantity of isolated DNA, only a fraction of the sequencing input molecules will be of a sufficient length (i.e., >10 kb) for which long-read sequencing can be beneficial. Issues of input DNA quantity can be addressed by inclusion of clever amplification approaches; however, amplification can also be a source of significant bias, hampering the utility of sequencing for quantitative ecological studies (Marine et al., 2014).

By and large, the scientific focus of metagenomic studies has been understanding the genetic and population diversity and community composition of the virioplankton. However, less attention has been paid to leveraging metagenomic data for predicting the prevailing life history characteristics of virioplankton populations. The individual life history traits of virus–host systems can critically shape carbon fate within oceanic ecosystems, for example, pelagiphages and EhVs capable of toggling between temperate and virulent life cycles. This lack of attention largely stems from an inability to accurately predict the phenotypes of an unknown virus based on its genome sequence alone. Nevertheless, there are promising signs that environmental virologists will eventually overcome the genotype to phenotype knowledge gap and more fully leverage metagenomic information for making informed predictions of how the collection of infection phenotypes observed across virioplankton populations may impact carbon flow within oceanic microbial communities. For example, the presence of auxiliary metabolic genes within specific virioplankton populations demonstrates potential infection impacts on the nitrogen (Ahlgren et al., 2019; Sullivan et al., 2010; Zhao et al., 2022), sulfur (Anantharaman et al., 2014; Zhao et al., 2022), and phosphorus (Huang et al., 2021; Kathuria and Martiny, 2011; Kelly et al., 2013) cycles, as well as host photosynthesis (Bailey et al., 2004; Fridman et al., 2017; Lindell et al., 2005; Mann et al., 2003). Recent work has hypothesized that specific single amino acid changes in PolA indicate whether an unknown phage more likely follows a lytic, or lysogenic/pseudolysogenic life history strategy (Schmidt et al., 2014). Examining genes neighboring PolA within the replication module such as helicase or RNR can provide even greater resolution as to the potential infection dynamics traits of an unknown virus (Nasko et al., 2018; Sakowski et al., 2014). While so many of the putative genes identified within viral genomes show no known function through sequence homology, genome replication genes are well known, common, and widely distributed across evolutionarily distant viral populations, making these genes good targets for building hypotheses connecting viral genotype to phenotype. Refining and testing genome to phenome hypotheses will require both reductionist study of model oceanic virus–host systems and application of metagenomics approaches for observing the behavior of specific viral populations in the ocean. The "holy grail" of this work will be in developing analytical tools based on validated genome to phenome linkages that predict the infection dynamics phenotypes of the hundreds of virus–host pairs observed within marine microbial communities.

The ubiquitous presence and activity of viruses in microbial communities and their extraordinary genetic diversity, which includes the presence of central cellular metabolic genes (i.e., termed auxiliary metabolic genes) in viral genomes, has fundamentally changed scientific perspectives of viral impacts on the life history of cellular microbes. Many now acknowledge that a virocell is fundamentally different from an uninfected cell in ways that extend beyond simply its physiological trajectory for making new viruses and eventual death by lysis (Forterre, 2011; Rosenwasser et al., 2016; Zimmerman et al., 2020). However, the scientific challenges in understanding the unique nature of the virocell lie first in observing the differences between infected and uninfected cells and then interpreting how these differences may impact ecosystem processes. With regards to the first challenge, the chemical constitution of a cell or its surrounding environment can be observed with high precision and sensitivity using metabolomics. Analyses of metabolites within *Pseudomonas aeruginosa* cells infected with one of six phylogenetically distinct phages demonstrated that metabolite alterations were clearly phage-specific, falling along a continuum from exhausting all existing cellular resources for phage production to modulating cellular metabolism for new resource production (De Smet et al., 2016). Not surprisingly, many of the metabolic changes occurring in infected cells are geared towards increasing the production or availability of nucleotides for viral genome replication (De Smet et al., 2016; Howard-Varona et al., 2020; Hurwitz et al., 2014). A study using a combination of proteomic (i.e., 'omic analysis of a complex collection of proteins) and transcriptomic analyses of a marine *Pseudoalteromonas* species infected with one of two different lytic phages came to a similar conclusion that each phage differentially altered host metabolism (Howard-Varona et al., 2020). Intriguingly, observed differences in cellular metabolism reflected phage fitness and its degree of complementarity with the host's codon usage patterns. The phage with lower complementarity required greater metabolic effort and resources from the host cell in producing phage particles. These metabolic differences could subsequently translate into differential effects on the composition of DOM released from lysis.

Metabolomic investigations have confirmed that viral infection changes the character of DOM exuded or released from lysed bacteria. A study of marine roseobacter species *Sulfitobacter* sp. 2047 found that, like *Pseudoalteromonas,* phage infection redirected most of cellular nutrients into phage production (Ankrah et al., 2014b). Surprisingly, most of the extracellular small metabolite compounds existing in uninfected control cultures showed reduced concentration with viral infection. The authors concluded that surviving uninfected cells rapidly consumed newly available small metabolite compounds confirming earlier work hypothesizing such DOM exchanges between infected and uninfected cells (Middelboe and Lyck, 2002; Middelboe et al., 1996). A study observing thousands of compounds within the DOM of infected and uninfected marine *Synechococcus* WH7803 cultures found that DOM released from infected cells was substantially more complex and starkly differed from the collection of DOM compounds exuded from cells in uninfected control cultures or mechanically lysed cells (Ma et al., 2018). In particular, virus-induced DOM (vDOM) was enriched in peptides resulting from the protolysis of *Synechococcus'* major light harvesting protein, phycoerythrin, making infected cells an important source of high molecular weight nitrogen compounds within the DOM. vDOM released through viral lysis of picocyanobacteria, such as *Synechococcus*, is particularly important in the oceanic carbon cycle given the importance of this host group in the global ocean. Recent metabolomics work has demonstrated that vDOM released from picocyanobacterial infection is readily processed by heterotrophic bacterioplankton fueling increases in the diversity and complexity of these communities (Zhao et al., 2019a).

While these metabolomic studies have confirmed a unique role for vDOM in the oceanic carbon cycle, they may also provide new tools for examining the carbon cycle impacts of viral infection. By connecting specific DOM compounds with specific DOM release mechanisms (e.g., exudation or viral lysis), metabolomics may provide specific biomarkers of viral infection within oceanic microbial communities. Perhaps the best example for the use of chemical biomarkers in tracking infection comes from the previously mentioned *E. huxleyi*–EhV system. Sequencing revealed that the EhV genome encodes an entire glycosphingolipid synthesis pathway (Wilson et al., 2005). Subsequent metabolomic analysis confirmed that indeed a unique viral glycosphingolipid (vGSL) was synthesized by this pathway and incorporated into EhV virions (Vardi et al., 2009). The virus likely uses its unique vGSL for stimulating the programmed cell death response in *E. huxleyi* which ultimately results in cell lysis and release of EhV. Field investigations leveraged these unique host and viral lipid biomarkers for diagnosing *in situ* levels of viral infection within natural coccolithophore blooms detected through satellite remote sensing (Laber et al., 2018). Biomarker-based characterization showed that EhV infection significantly enhanced biological pump processes with greater levels of OM aggregation and downward flux of particulate organic and inorganic carbon. These effects mostly occurred during early bloom infection when sinking material was notably enriched in infected *E. huxleyi* cells. It is unlikely that the *E. huxleyi*–EhV system is unique in producing chemical biomarkers capable of demonstrating *in situ* infection levels. The future looks bright for discovering new means for quantifying viral infection impacts on the oceanic carbon cycle through application of the metabolomic tool box to investigations of virus–host interactions in the sea.

4.2 Incorporating viral processes into global marine carbon cycling models

Reliably estimating the past, present, and future of the carbon cycle necessarily requires including the multi-layered role of viruses in the dynamics of different nutrient cycles. The benefits of accounting for viruses in models for oceanic biogeochemistry go beyond greater quantitative accuracy. Including virus–host dynamics in models modifies existing pathways of biogeochemical transformation and adds mechanisms to the description of the marine food web (e.g., viral shuttle/shunt, see above and Collins et al., 2015; Lehahn et al., 2014). Biogeochemical transformation pathways are important in quantitatively describing and predicting not only the flux of carbon and other nutrients but also their timing, and therefore are essential to the understanding of the dynamics of blooms, succession, and ultimately the biogeography of phytoplankton and how it changes over time.

Importantly, changes in the spatio-temporal distribution of phytoplankton and bacteria ripple across the rest of the marine food web, thus affecting estimates of zooplankton density and community composition, which in turn affects the nekton (fish and other microorganisms that swim). Additionally, improved models can create testable predictions and hypotheses for more targeted field, laboratory, and metagenomic research, or consider "what-if" scenarios (e.g., different climate change scenarios) at spatio-temporal scales that are not attainable empirically.

Only recently, however, have viruses started to be explicitly accounted for in models that aim to predict primary productivity (reviewed in Mateus, 2017). Neglected in most oceanic biogeochemistry models, attempts to add viruses have relied on indirect effective mortality terms for the phytoplankton dynamic equations (e.g., density-dependent terms) (Baretta-Bekker et al., 1995; Beltrami and Carroll, 1994; Chattopadhyay and Pal, 2002; Hasumi and Nagata, 2014; Singh et al., 2004; Stock et al., 2020). Only now do ecosystem models consider viruses explicitly as dynamic agents through independent equations, although still only at a local/regional level (i.e., no global descriptions) (Béchette et al., 2013; Keller and Hood, 2011, 2013; Richards, 2017; Talmy et al., 2019; Weitz et al., 2015; Xie et al., 2022).

The incorporation of viral dynamics into biogeochemical models of any type is mostly hindered by the multiple scales at which viruses shape such cycles. On one hand, experimental and laboratory information is sufficiently detailed for building and validating models pertaining to the individual virocell or to population levels under controlled conditions. However, such scales are too fine-grained (and too ideal) compared with the typical spatio-temporal scales on which global biogeochemical models focus (Follows et al., 2007; Pahlow et al., 2020; Stock et al., 2020). On the other hand, representing the vast diversity of potential hosts (Bonachela et al., 2016; Finkel et al., 2010) and viruses (Breitbart, 2012; Brussaard, 2004) within models is impossible and, despite the specificity of viral infection which limits the possible virus–host combinations (Poullain et al., 2008), we lack information on which pairings are eventually realized (Kauffman et al., 2022).

Most of the information used for developing and parametrizing virus–host models is obtained in controlled environments, as such conditions facilitate the focus on specific aspects of the virus–host interaction while avoiding confounding effects. Such conditions can, however, also lead to overly idealistic perceptions of the reality of virus–host dynamics. For example, typical infection experiments start with a very abundant host population that grows at its maximum growth rate, ensuring infection of a large proportion of hosts (Hadas et al., 1997). In the ocean, however, those conditions are the exception more than the rule, and considering more realistic scenarios have led to observations that have challenged our intuition about virus–host interactions. Infection experiments with *E. huxleyi* and EhV using initial host densities similar to those observed in the field (orders of magnitude lower than those typically used in the lab) have described a temperance life history for such viruses, which previous experiments consistently reported as purely lytic (Section 3.2.1 and Knowles et al., 2020 for details). In these cases, viral infection did not result in lysis until the host population reached high densities and hosts started to show physiological stress, leading to the hypothesis, supported through experimental and modeling data, that the lytic switch was triggered by host physiological changes. Similarly, experiments measuring viral traits and performance when the host was not under ideal growth conditions have reported values that are very different from those typically used in models (Cheng et al., 2015; Golec et al., 2014; Hadas et al., 1997; Kranzler et al., 2019; Maat et al., 2016; Piedade et al., 2018; Van Etten et al., 1983; You et al., 2002). For example, for the bacterium *E. coli* and its infecting T phage, a higher host growth rate resulted in shorter infections that produced more numerous offspring (i.e., shorter latent periods and larger burst sizes) (Golec et al., 2014; Hadas et al., 1997; You et al., 2002). Increased temperature had a similar effect for viruses infecting the chlorophyte *Micromonas polaris* (Maat et al., 2017; Piedade et al., 2018). Similar qualitative effects have also been shown in green algae and diatoms for high versus low nutrient availability and for high versus low light intensity (Bratbak et al., 1998a; Cheng et al., 2015; Kranzler et al., 2019; Piedade et al., 2018).

When included in models, this dependence of viral traits on host physiology (termed "viral plasticity" since the host cell constitutes the virus' reproductive environment) leads to ecological and evolutionary predictions very different from those built with standard models that use "ideal" viral trait values (Bonachela et al., 2022; Choua and Bonachela, 2019; Choua et al., 2020; Edwards and Steward, 2018). For example, burst size is expected to decrease as host growth conditions worsen (see above, and Webb et al., 1982), as opposed to the fixed ideal value used in standard models. There are other reasons why the burst size can show discrepancies between values estimated under ideal conditions and open-ocean conditions. Key assumptions in the standard calculation of the burst size, for example, may not be applicable to all virus–host systems. A typical way to calculate burst size is by monitoring the growth of the host and extracellular virus populations after an initial adsorption event, and dividing the change in viral density by the decline in host density. The assumption is that all new extracellular viruses result from lytic events, which in turn (it is assumed) are the only reason host cells die. Even under ideal conditions, however, some viruses like EhV can show initial replication through budding (Mackinder et al., 2009), which constitutes viral production without lysis thus tainting the calculation of the burst size. A solution would be to monitor intracellular viruses throughout an infection cycle. In systems such as *E. coli* and T viruses, this can be done by regularly sampling the host population during one infection cycle, artificially bursting open cells for each sample measuring the number of infective (i.e., mature) viruses, and finally dividing that number by the number of hosts that have been lysed (e.g., You et al., 2002). In the case of phytoplankton, however, the longer timescales for host and viral growth necessitate less labor-intensive solutions such as viral staining and qPCR techniques (Knowles et al., 2020). Incidentally, monitoring intracellular viruses would bring the estimate of the burst size closer to the definition used in theory (number of new mature viruses released per lytic event).

The values typically used for the adsorption rate constitute another remarkable example of divergences between estimates for ideal, controlled environments versus realistic ones. The adsorption rate is defined as the rate at which an individual virus encounters a host cell and attaches to it successfully, using its specific host receptor (Poullain et al., 2008). Estimates obtained in laboratory experiments and used in virus–host models (e.g., De Paepe and Taddei, 2006; Weitz et al., 2005) are much larger than the predictions that physical models make assuming randomly diffusing particles under ocean-like conditions (e.g., Knowles et al., 2020; Murray and Jackson, 1992). The solution to this apparent discrepancy may be, again, taking into consideration the diversity of hosts and viruses when estimating the adsorption rate, and that their spatio-temporal distribution and growth conditions are far from homogeneous or ideal (Kauffman et al., 2022). Understanding the origin of this quantitative gap between ideal and realistic values for burst size and adsorption rate is important because any virus–host infection dynamics model includes these two parameters/traits as part of the infection term.

Importantly, the much lower rates at which host cells and viruses are predicted to meet under realistic scenarios, and the lower productivity of those encounters, somewhat throws into question that viruses are actually responsible for the high levels of mortality assumed for the marine microbial community. Nonetheless, reported mortality ranges are admittedly quite broad, from 5 to 15% for some cyanobacteria to up to 50% of *E. huxleyi* cells during bloom demise, and always larger when considering controlled experiments, e.g., mortality rates up to 100% for *E. huxleyi* in mesocosm experimental blooms (reviewed in Fuhrman, 1999; Zimmerman et al., 2020). This mismatch between theoretically assumed and empirically measured mortality is reflected in what, for years, has been the preferred method for estimating viral mortality for phytoplankton: viral dilution experiments. Originally devised to estimate mortality due to grazing (Landry and Hassett, 1982), the idea underlying the experiment is that dilution of grazers and phytoplankton reduces grazing rates and thus increases phytoplankton net growth rate with respect to undiluted samples. Changes in apparent growth rate (AGR) with dilution (i.e., slope of AGR) enable the calculation of the grazing rate. The original methodology, and calculation of AGRs and thus grazing rates, was devised using theoretical arguments that assumed, for example, a duration of the experiment such that the population of grazers remained constant (24 h). The methodology was later adapted to measure phytoplankton viral mortality by adding an additional step in which viruses were also diluted (Evans et al., 2003); however, the theoretical assumptions (some of which were already questionable for some grazers (Dolan and McKeon, 2005; Evans and Paranjape, 1992) were not adapted to viruses, for which some of these assumptions further break down (e.g., population of viruses remaining constant during the 24 h of the experiment). Unknowingly, experiments using the dilution technique disregarded as artifacts or outliers any data that produced unusual predictions (e.g., opposite slope for AGR), thus missing valuable information about the top-down pressure occurring on the focal phytoplankton population (Baudoux et al., 2006; Dolan and McKeon, 2005). Recent efforts integrating the ecological aspects of the dilution experiment and theory have proposed revisions to the methodology (e.g., measuring not only phytoplankton abundance but also grazer and viral densities, as is routinely done in bacterioplankton dilution experiments) that allow for a reinterpretation and use of the data collected with it (Beckett and Weitz, 2017, 2018; Talmy et al., 2019).

These examples illustrate how the dialogue across disciplines and methodologies can help improve both empirical methods and theoretical models. Further, such efforts may be the key to unlocking the inclusion of viral dynamics into biogeochemical models. For example, past and current work to understand how viral traits depend on host physiology will provide key information for models to move from unrealistic fixed viral trait values to values that change in space and time with host growth conditions.

Inclusion of viral plasticity in models will improve realism and predictability, and also may help ameliorate the diversity problem as viral plasticity introduces variability and thus a better representation of the viral trait space in models. Trait-based approaches commonly used to model, for example, phytoplankton diversity (Bonachela et al., 2016; Follows and Dutkiewicz, 2011; Litchman and Klausmeier, 2008; Schmidt, 2019), are, for that reason, more and more used in biogeochemical models (Follows et al., 2007) and may offer a computationally cost-effective level of detail for representing viral diversity. Such approaches can bridge the gap between controlled experiments and large-scale predictions by providing an opportunity to build multi-layered models and study increasingly complex communities or scenarios. For example, data obtained in controlled experiments for isolated virus–host systems can be used to inform a model in which such systems are brought together to form a specific, documented, diverse community (which may or may not be realizable in an experimental setup), with the aim of predicting the dynamics of the community as a whole. Such an approach has been used to, for example, estimate the competing effects of grazing and viral mortality (Talmy et al., 2019) and top-down versus bottom-up regulation (Pourtois et al., 2020; Weitz et al., 2015) on primary production and carbon export, steps that can be used as building blocks to the inclusion of viruses in global models for oceanic biogeochemistry.

A possible argument against using trait-based approaches for representing virioplankton is the intertwinement between host and virus within the virocell, which makes the definition of "viral traits" a delicate matter (DeLong et al., 2022). As an alternative, a more detailed model at the virocell level can be scaled-up to much larger spatio-temporal scales using theoretical tools from disciplines such as statistical physics (Gardiner, 2009; van Kampen, 2007), which have been adapted to the ecological context repeatedly (Flierl et al., 1999; Levin, 1992). The end product would be equations that, keeping a discernible link with the microscopic level, describe virus–host dynamics at a coarser scale that is suitable for global models.

A last important piece of information required for representing realistic levels of diversity pertains to the virus–host pairs that are realized (i.e., actually interacting) in the ocean. A naive approach would consider that all hosts are infected by all viruses; however, marine viruses typically show a narrow host range and hosts show remarkable resistance levels (Kauffman et al., 2022). Although an important body of work has focused on pairwise interactions (one host and one virus population), increasingly empirical and theoretical research uses the tools of network theory and statistical mechanics to describe the interactions of whole communities (Beckett and Williams, 2013; Flores et al., 2011, 2013; Kauffman et al., 2022; Moebus and Nattkemper, 1981). These analyses have revealed complex, multi-scale structures for the interaction network that show clusters at one scale (modules) and, within such clusters, nestedness and submodules. Including viruses in global models will require documenting and describing empirically, and being able to predict theoretically, not only the structure of the network but also how it changes over time.

5. Overall takeaways and conclusions

Climate change poses an immediate and long-term threat to the ocean's current structure and function, both biogeochemically and ecologically. Currently, it is unclear how changes in oceanic temperature, circulation, stratification, and acidification affect whole microbial communities, their associated viromes, and consequential carbon flux. Untangling the physical, biological, and chemical mechanisms that shape virus-mediated carbon cycling provides opportunities for rebuilding integrated frameworks, experimental approaches, and models in a more holistic way that considers ocean-relevant spatial and temporal scales.

As this review has shown, achieving this goal will require research collaboration that pushes the boundaries of microbial and viral research by integrating model systems, experimental work, and field microbial ecology investigations alongside rigorous consideration of prevailing biogeochemical and physical contexts, all of which shape the probability, outcome, and consequences of virus–host interactions in the sea. For example, several virus–host model systems critical to light-derived carbon cycling dynamically respond to changes in environmental context. Thus, these dynamic physiological changes and viral responses must be considered in empirical and theoretical investigations.

Given that rapid evolution is a hallmark of both microbes and viruses, biogeochemical changes and concomitant host physiological changes could alter evolutionary trajectories of both viruses and hosts. An unpredictable environment will favor a generalist's capacity to plastically respond to, navigate, and survive various conditions over a specialist's more rigid phenotypes (Huey and Slatkin, 1976; Hughes et al., 2007; Levins, 1968), even if the specialist is able to outcompete the generalist under stable environmental conditions (Hughes et al., 2007). As plasticity is a trait unto itself that confers improved fitness in fluctuating or variable environments

(Kassen, 2002; Levins, 1968; von Meijenfeldt et al., 2022), direct tests of microbial evolution with elevated CO_2 partial pressures (pCO_2) found that fluctuating environments selected for phenotypic plasticity in *Ostreococcus* lineages (an important oceanic chlorophyte) (Schaum and Collins, 2014). In contrast, under constant high pCO_2 conditions, lineages evolved directional tolerance at the cost of phenotypic plasticity such that populations failed to survive in their ancestral conditions (Schaum and Collins, 2014). Notably, how interactions respond to the cascading effects of evolving host phenotypic plasticity under fluctuating environments remains an unexplored question.

In addition, steadily closing the genome-to-phenome knowledge gap for marine viruses (and their microbial hosts) will unlock the extraordinary observational power of 'omics tools and approaches for examining hypothesized responses of oceanic microbial communities to global climate change. Omics-driven observations will also help in unveiling the ecological and evolutionary factors influencing the assembly and dynamics of virus–host interaction networks, and thus contribute to developing predictive models capturing such dynamics.

Numerous challenges remain that prevent well constrained and explicit inclusion of viruses in the description and prediction of global biogeochemical cycles. As illustrated here, however, interdisciplinary dialogue that allows for the combination of empirical and theoretical expertise across fields is helping in the search for solutions that will fully incorporate viral dynamics into our understanding of the global ocean, the virus' dominion.

Acknowledgments

HL, BDF, and KEW were supported by the National Science Foundation (OIA-1736030). KDB and CTJ were supported by the National Science Foundation (OIA-2021032; OCE-1537951; OCE-1061883), NASA (15-RRNES15-0011; 0NSSC18K1563; FINESST 80NSSC19K1299), and the Gordon and Betty Moore Foundation (award #3789). KT was supported by the National Science Foundation (OCE-2049386). JAB was supported by the Simons Foundation (award #82610).

References

Ahlgren, N.A., Rocap, G., 2006. Culture isolation and culture-independent clone libraries reveal new marine *Synechococcus* ecotypes with distinctive light and N physiologies. Appl. Environ. Microbiol. 72 (11), 7193–7204.

Ahlgren, N.A., Ren, J., Lu, Y.Y., Fuhrman, J.A., Sun, F., 2017. Alignment-free oligonucleotide frequency dissimilarity measure improves prediction of hosts from metagenomically-derived viral sequences. Nucleic Acids Res. 45 (1), 39–53.

Ahlgren, N.A., Fuchsman, C.A., Rocap, G., Fuhrman, J.A., 2019. Discovery of several novel, widespread, and ecologically distinct marine *Thaumarchaeota* viruses that encode amoC nitrification genes. ISME J. 13 (3), 618–631.

Allen, M.J., Forster, T., Schroeder, D.C., Hall, M., Roy, D., Ghazal, P., Wilson, W.H., 2006a. Locus-specific gene expression pattern suggests a unique propagation strategy for a giant algal virus. J. Virol. 80 (15), 7699–7705.

Allen, M.J., Schroeder, D.C., Wilson, W.H., 2006b. Preliminary characterisation of repeat families in the genome of EhV-86, a giant algal virus that infects the marine microalga *Emiliania huxleyi*. Arch. Virol. 151 (3), 525–535.

Alonso-Sáez, L., Morán, X.A.G., Clokie, M.R., 2018. Low activity of lytic pelagiphages in coastal marine waters. ISME J. 12 (8), 2100–2102.

Anantharaman, K., Duhaime, M.B., Breier, J.A., Wendt, K.A., Toner, B.M., Dick, G.J., 2014. Sulfur oxidation genes in diverse deep-sea viruses. Science 344 (6185), 757–760.

Ankrah, N.Y., Budinoff, C.R., Wilson, W.H., Wilhelm, S.W., Buchan, A., 2014a. Genome sequences of two temperate phages, ΦCB2047-A and ΦCB2047-C, infecting Sulfitobacter sp. strain 2047. Genome Announcements 2 (3), e00108–e00114.

Ankrah, N.Y.D., May, A.L., Middleton, J.L., Jones, D.R., Hadden, M.K., Gooding, J.R., LeCleir, G.R., Wilhelm, S.W., Campagna, S.R., Buchan, A., 2014b. Phage infection of an environmentally relevant marine bacterium alters host metabolism and lysate composition. ISME J. 8 (5), 1089–1100.

Armengol, L., Calbet, A., Franchy, G., Rodríguez-Santos, A., Hernández-León, S., 2019. Planktonic food web structure and trophic transfer efficiency along a productivity gradient in the tropical and subtropical Atlantic Ocean. Sci. Rep. 9 (1), 1–19.

Armstrong, R.A., Lee, C., Hedges, J.I., Honjo, S., Wakeham, S.G., 2001. A new, mechanistic model for organic carbon fluxes in the ocean based on the quantitative association of POC with ballast minerals. Deep-Sea Res. II Top. Stud. Oceanogr. 49 (1-3), 219–236.

Arsenieff, L., Kimura, K., Kranzler, C.F., Baudoux, A.C., Thamatrakoln, K., 2022. Diatom Viruses. In: The Molecular Life of Diatoms. Springer, Cham, pp. 713–740.

Bachy, C., Charlesworth, C.J., Chan, A.M., Finke, J.F., Wong, C.H., Wei, C.L., Sudek, S., Coleman, M.L., Suttle, C.A., Worden, A.Z., 2018. Transcriptional responses of the marine green alga *Micromonas pusilla* and an infecting prasinovirus under different phosphate conditions. Environ. Microbiol. 20 (8), 2898–2912.

Bailey, S., Clokie, M.R., Millard, A., Mann, N.H., 2004. Cyanophage infection and photoinhibition in marine cyanobacteria. Res. Microbiol. 155 (9), 720–725.

Balch, W., Drapeau, D., Bowler, B., Booth, E., 2007. Prediction of pelagic calcification rates using satellite measurements. Deep-Sea Res. II Top. Stud. Oceanogr. 54 (5-7), 478–495.

Baretta-Bekker, J.G., Baretta, J.W., Rasmussen, E.K., 1995. The microbial food web in the European regional seas ecosystem model. Neth. J. Sea Res. 33 (3-4), 363–379.

Barton, S., Yvon-Durocher, G., 2019. Quantifying the temperature dependence of growth rate in marine phytoplankton within and across species. Limnol. Oceanogr. 64 (5), 2081–2091.

Basterretxea, G., Font-Munoz, J.S., Tuval, I., 2020. Phytoplankton orientation in a turbulent ocean: a microscale perspective. Front. Mar. Sci. 7, 185.

Batchelder, H.P., Edwards, C.A., Powell, T.M., 2002. Individual-based models of copepod populations in coastal upwelling regions: implications of physiologically and environmentally influenced diel vertical migration on demographic success and nearshore retention. Prog. Oceanogr. 53 (2-4), 307–333.

Bates, N.R., Mathis, J.T., 2009. The Arctic Ocean marine carbon cycle: evaluation of air-sea CO 2 exchanges, ocean acidification impacts and potential feedbacks. Biogeosciences 6 (11), 2433–2459.

Baudoux, A.C., Brussaard, C.P., 2005. Characterization of different viruses infecting the marine harmful algal bloom species *Phaeocystis globosa*. Virology 341 (1), 80–90.

Baudoux, A.C., Brussaard, C.P., 2008. Influence of irradiance on virus–algal host interactions 1. J. Phycol. 44 (4), 902–908.

Baudoux, A.C., Noordeloos, A.A., Veldhuis, M.J., Brussaard, C.P., 2006. Virally induced mortality of *Phaeocystis globosa* during two spring blooms in temperate coastal waters. Aquat. Microb. Ecol. 44 (3), 207–217.

Beaulaurier, J., Luo, E., Eppley, J.M., Den Uyl, P., Dai, X., Burger, A., Turner, D.J., Pendelton, M., Juul, S., Harrington, E., DeLong, E.F., 2020. Assembly-free single-molecule sequencing recovers complete virus genomes from natural microbial communities. Genome Res. 30 (3), 437–446.

Béchette, A., Stojsavljevic, T., Tessmer, M., Berges, J.A., Pinter, G.A., Young, E.B., 2013. Mathematical modeling of bacteria–virus interactions in Lake Michigan incorporating phosphorus content. J. Great Lakes Res. 39 (4), 646–654.

Becker, J.W., Berube, P.M., Follett, C.L., Waterbury, J.B., Chisholm, S.W., DeLong, E.F., Repeta, D.J., 2014. Closely related phytoplankton species produce similar suites of dissolved organic matter. Front. Microbiol. 5, 111.

Beckett, S.J., Weitz, J.S., 2017. Disentangling niche competition from grazing mortality in phytoplankton dilution experiments. PLoS One 12 (5), e0177517.

Beckett, S.J., Weitz, J.S., 2018. The effect of strain level diversity on robust inference of virus-induced mortality of phytoplankton. Front. Microbiol. 9, 1850.

Beckett, S.J., Williams, H.T., 2013. Coevolutionary diversification creates nested-modular structure in phage–bacteria interaction networks. Interface Focus 3 (6), 20130033.

Behrenfeld, M.J., Boss, E.S., 2018. Student's tutorial on bloom hypotheses in the context of phytoplankton annual cycles. Glob. Chang. Biol. 24 (1), 55–77.

Behrenfeld, M.J., Moore, R.H., Hostetler, C.A., Graff, J., Gaube, P., Russell, L.M., Chen, G., Doney, S.C., Giovannoni, S., Liu, H., Proctor, C., 2019. The North Atlantic aerosol and marine ecosystem study (NAAMES): science motive and mission overview. Front. Mar. Sci. 6, 122.

Beltrami, E., Carroll, T.O., 1994. Modeling the role of viral disease in recurrent phytoplankton blooms. J. Math. Biol. 32 (8), 857–863.

Berelson, W.M., Balch, W.M., Najjar, R., Feely, R.A., Sabine, C., Lee, K., 2007. Relating estimates of CaCO3 production, export, and dissolution in the water column to measurements of CaCO3 rain into sediment traps and dissolution on the sea floor: A revised global carbonate budget. Glob. Biogeochem. Cycles 21 (1).

Bergh, Ø., Børsheim, K.Y., Bratbak, G., Heldal, M., 1989. High abundance of viruses found in aquatic environments. Nature 340 (6233), 467–468.

Bidle, K.D., 2015. The molecular ecophysiology of programmed cell death in marine phytoplankton. Annu. Rev. Mar. Sci. 7, 341–375.

Bidle, K.D., Kwityn, C.J., 2012. Assessing the role of caspase activity and metacaspase expression on viral susceptibility of the coccolithophore, *Emiliania huxleyi* (Haptophyta). J. Phycol. 48 (5), 1079–1089.

Bidle, K.D., Vardi, A., 2011. A chemical arms race at sea mediates algal host–virus interactions. Curr. Opin. Microbiol. 14 (4), 449–457.

Bidle, K.D., Haramaty, L., Barcelos e Ramos, J. and Falkowski, P., 2007. Viral activation and recruitment of metacaspases in the unicellular coccolithophore, *Emiliania huxleyi*. Proc. Natl. Acad. Sci. 104 (14), 6049–6054.

Bin Jang, H., Bolduc, B., Zablocki, O., Kuhn, J.H., Roux, S., Adriaenssens, E.M., Brister, J.R., Kropinski, A.M., Krupovic, M., Lavigne, R., Turner, D., 2019. Taxonomic assignment of uncultivated prokaryotic virus genomes is enabled by gene-sharing networks. Nat. Biotechnol. 37 (6), 632–639.

Bindoff, N.L., Cheung, W.W., Kairo, J.G., Arístegui, J., Guinder, V.A., Hallberg, R., Hilmi, N.J.M., Jiao, N., Karim, M.S., Levin, L., O'Donoghue, S., 2019. Changing ocean, marine ecosystems, and dependent communities. In: IPCC Special Report on the Ocean and Cryosphere in a Changing Climate, pp. 477–587.

Bischoff, V., Bunk, B., Meier-Kolthoff, J.P., Spröer, C., Poehlein, A., Dogs, M., Nguyen, M., Petersen, J., Daniel, R., Overmann, J., Göker, M., Simon, M., Brinkhoff, T., Moraru, C., 2019. Cobaviruses—a new globally distributed phage group infecting Rhodobacteraceae in marine ecosystems. ISME J. 13 (6), 1404–1421.

Boisvert, S., Raymond, F., Godzaridis, É., Laviolette, F., Corbeil, J., 2012. Ray Meta: scalable de novo metagenome assembly and profiling. Genome Biol. 13 (12), 1–13.

Bolaños, L.M., Choi, C.J., Worden, A.Z., Baetge, N., Carlson, C.A., Giovannoni, S., 2021. Seasonality of the microbial community composition in the North Atlantic. Front. Mar. Sci. 8, 624164.

Bonachela, J.A., Klausmeier, C.A., Edwards, K.F., Litchman, E., Levin, S.A., 2016. The role of phytoplankton diversity in the emergent oceanic stoichiometry. J. Plankton Res. 38 (4), 1021–1035.

Bonachela, J.A., Choua, M., Heath, M.R., 2022. Unconstrained coevolution of bacterial size and the latent period of plastic phage. PLoS One 17 (5), e0268596.

Brandenburg, K.M., Velthuis, M., Van de Waal, D.B., 2019. Meta-analysis reveals enhanced growth of marine harmful algae from temperate regions with warming and elevated CO2 levels. Glob. Chang. Biol. 25 (8), 2607–2618.

Bratbak, G., Egge, J.K., Heldal, M., 1993. Viral mortality of the marine alga *Emiliania huxleyi* (Haptophyceae) and termination of algal blooms. Mar. Ecol. Prog. Ser., 39–48.

Bratbak, G., Jacobsen, A., Heldal, M., 1998a. Viral lysis of *Phaeocystis pouchetii* and bacterial secondary production. Aquat. Microb. Ecol. 16 (1), 11–16.

Bratbak, G., Jacobsen, A., Heldal, M., Nagasaki, K., Thingstad, F., 1998b. Virus production in *Phaeocystis pouchetii* and its relation to host cell growth and nutrition. Aquat. Microb. Ecol. 16 (1), 1–9.

Breitbart, M., 2012. Marine viruses: truth or dare. Annu. Rev. Mar. Sci. 4 (1), 425–448.

Breitbart, M., Bonnain, C., Malki, K., Sawaya, N.A., 2018. Phage puppet masters of the marine microbial realm. Nat. Microbiol. 3 (7), 754–766.

Brinkmeyer, R., Knittel, K., Jürgens, J., Weyland, H., Amann, R., Helmke, E., 2003. Diversity and structure of bacterial communities in Arctic versus Antarctic pack ice. Appl. Environ. Microbiol. 69 (11), 6610–6619.

Broecker, W., Clark, E., 2009. Ratio of coccolith CaCO3 to foraminifera CaCO3 in late Holocene deep sea sediments. Paleoceanography 24 (3).

Brown, C.M., Bidle, K.D., et al., 2014. Attenuation of virus production at high multiplicities of infection in Aureococcus anophagefferens. Virology 466, 71–81.

Brown, C.M., Campbell, D.A., Lawrence, J.E., 2007. Resource dynamics during infection of *Micromonas pusilla* by virus MpV-Sp1. Environ. Microbiol. 9 (11), 2720–2727.

Brussaard, C.P., 2004. Viral control of phytoplankton populations—a review 1. J. Eukaryot. Microbiol. 51 (2), 125–138.

Brussaard, C.P.D., Riegman, R., Noordeloos, A.A.M., Cadée, G.C., Witte, H., Kop, A.J., Nieuwland, G., Van Duyl, F.C., Bak, R.P.M., 1995. Effects of grazing, sedimentation and phytoplankton cell lysis on the structure of a coastal pelagic food web. Mar. Ecol. Prog. Ser. 123, 259–271.

Brussaard, C.P.D., Gast, G.J., Van Duyl, F.C., Riegman, R., 1996. Impact of phytoplankton bloom magnitude on a pelagic microbial food web. Mar. Ecol. Prog. Ser. 144, 211–221.

Brussaard, C.P., Thyrhaug, R., Marie, D., Bratbak, G., 1999. Flow cytometric analyses of viral infection in two marine phytoplankton species, *Micromonas pusilla* (Prasinophyceae) and *Phaeocystis pouchetii* (Prymnesiophyceae). J. Phycol. 35 (5), 941–948.

Brussaard, C.P., Marie, D., Thyrhaug, R., Bratbak, G., 2001. Flow cytometric analysis of phytoplankton viability following viral infection. Aquat. Microb. Ecol. 26 (2), 157–166.

Brussaard, C.P.D., Short, S.M., Frederickson, C.M., Suttle, C.A., 2004. Isolation and phylogenetic analysis of novel viruses infecting the phytoplankton *Phaeocystis globosa* (Prymnesiophyceae). Appl. Environ. Microbiol. 70 (6), 3700–3705.

Brussaard, C.P.D., Kuipers, B., Veldhuis, M.J.W., 2005a. A mesocosm study of *Phaeocystis globosa* population dynamics: I. Regulatory role of viruses in bloom control. Harmful Algae 4 (5), 859–874.

Brussaard, C.P.D., Mari, X., Van Bleijswijk, J.D.L., Veldhuis, M.J.W., 2005b. A mesocosm study of *Phaeocystis globosa* (Prymnesiophyceae) population dynamics: II. Significance for the microbial community. Harmful Algae 4 (5), 875–893.

Brussaard, C.P., Bratbak, G., Baudoux, A.C., Ruardij, P., 2007. Phaeocystis and its interaction with viruses. In: Phaeocystis, major Link in the Biogeochemical Cycling of Climate-Relevant Elements. Springer, Dordrecht, pp. 201–215.

Brzezinski, M.A., Olson, R.J., Chisholm, S.W., 1990. Silicon availability and cell-cycle progression in marine diatoms. Mar. Ecol. Prog. Ser., 83–96.

Buchan, A., González, J.M., Moran, M.A., 2005. Overview of the marine Roseobacter lineage. Appl. Environ. Microbiol. 71 (10), 5665–5677.

Buchholz, H.H., Michelsen, M.L., Bolaños, L.M., Browne, E., Allen, M.J., Temperton, B., 2021a. Efficient dilution-to-extinction isolation of novel virus–host model systems for fastidious heterotrophic bacteria. ISME J. 15 (6), 1585–1598.

Buchholz, H.H., Michelsen, M., Parsons, R.J., Bates, N.R., Temperton, B., 2021b. Draft genome sequences of pelagimyophage Mosig EXVC030M and pelagipodophage Lederberg EXVC029P, isolated from Devil's Hole, Bermuda. Microbiol. Resour. Announcements 10 (7). e01325-20.

Burd, A.B., Jackson, G.A., 2009. Particle aggregation. Annu. Rev. Mar. Sci. 1, 65–90.

Caesar, L., McCarthy, G., Thornalley, D., Rahmstorf, S., 2020. The evolution of the Atlantic meridional overturning circulation for more than 1000 years. In: AGU Fall Meeting Abstracts. 2020. 038-02.

Cai, L., Ma, R., Chen, H., Yang, Y., Jiao, N., Zhang, R., 2019. A newly isolated roseophage represents a distinct member of Siphoviridae family. Virol. J. 16 (1), 1–9.

Campbell, L., Vaulot, D., 1993. Photosynthetic picoplankton community structure in the subtropical North Pacific Ocean near Hawaii (station ALOHA). Deep-Sea Res. I Oceanogr. Res. Pap. 40 (10), 2043–2060.

Carey, C.C., Ibelings, B.W., Hoffmann, E.P., Hamilton, D.P., Brookes, J.D., 2012. Eco-physiological adaptations that favour freshwater cyanobacteria in a changing climate. Water Res. 46 (5), 1394–1407.

Carlson, A.E., Clark, P.U., 2012. Ice sheet sources of sea level rise and freshwater discharge during the last deglaciation. Rev. Geophys. 50 (4).

Cassotta, S., Derksen, C., Ekaykin, A., Hollowed, A., Kofinas, G., MAckintosh, A., Melbourne-Thomas, J., Muelbert, M., Ottersen, G., Pritchard, H., Schuur, E.A.G., 2022. Special report on ocean and cryosphere in a changing chapter Intergovernmental Panel on Climate Change (IPCC). In: Chapter 3 Polar Issues. Cambridge University Press.

Castberg, T., Thyrhaug, R., Larsen, A., Sandaa, R.A., Heldal, M., Van Etten, J.L., Bratbak, G., 2002. Isolation and characterization of a virus that infects *Emiliania huxleyi* (Haptophyta). J. Phycol. 38 (4), 767–774.

Chan, Y.W., Mohr, R., Millard, A.D., Holmes, A.B., Larkum, A.W., Whitworth, A.L., Mann, N.H., Scanlan, D.J., Hess, W.R., Clokie, M.R., 2011. Discovery of cyanophage genomes which contain mitochondrial DNA polymerase. Mol. Biol. Evol. 28 (8), 2269–2274.

Chan, J.Z.M., Millard, A.D., Mann, N.H., Schäfer, H., 2014. Comparative genomics defines the core genome of the growing N4-like phage genus and identifies N4-like Roseophage specific genes. Front. Microbiol. 5, 506.

Chattopadhyay, J., Pal, S., 2002. Viral infection on phytoplankton–zooplankton system—a mathematical model. Ecol. Model. 151 (1), 15–28.

Chen, F., Lu, J., 2002. Genomic sequence and evolution of marine cyanophage P 60: a new insight on lytic and lysogenic phages. Appl. Environ. Microbiol. 68 (5), 2589–2594.

Chen, F., Wang, K., Stewart, J., Belas, R., 2006. Induction of multiple prophages from a marine bacterium: a genomic approach. Appl. Environ. Microbiol. 72 (7), 4995–5001.

Cheng, Y.S., Labavitch, J., VanderGheynst, J.S., 2015. Organic and inorganic nitrogen impact Chlorella variabilis productivity and host quality for viral production and cell lysis. Appl. Biochem. Biotechnol. 176 (2), 467–479.

Chénard, C., Wirth, J.F., Suttle, C.A., et al., 2016. Viruses infecting a freshwater filamentous cyanobacterium (*Nostoc* sp.) encode a functional CRISPR array and a proteobacterial DNA polymerase B. MBio 7 (3), e00067–16.

Cheng, K., Frenken, T., Brussaard, C.P., Van de Waal, D.B., 2019. Cyanophage propagation in the freshwater Cyanobacterium phormidium is constrained by phosphorus limitation and enhanced by elevated pCO2. Front. Microbiol. 10, 617.

Chevallereau, A., Pons, B.J., van Houte, S., Westra, E.R., 2022. Interactions between bacterial and phage communities in natural environments. Nat. Rev. Microbiol. 20 (1), 49–62.

Choua, M., Bonachela, J.A., 2019. Ecological and evolutionary consequences of viral plasticity. Am. Nat. 193 (3), 346–358.

Choua, M., Heath, M.R., Speirs, D.C., Bonachela, J.A., 2020. The effect of viral plasticity on the persistence of host-virus systems. J. Theor. Biol. 498, 110263.

Chu, T.C., Murray, S.R., Hsu, S.F., Vega, Q., Lee, L.H., 2011. Temperature-induced activation of freshwater Cyanophage AS-1 prophage. Acta Histochem. 113 (3), 294–299.

Clasen, J.L., Hanson, C.A., Ibrahim, Y., Weihe, C., Marston, M.F., Martiny, J.B., 2013. Diversity and temporal dynamics of Southern California coastal marine cyanophage isolates. Aquat. Microb. Ecol. 69 (1), 17–31.

Coleman, M.L., Sullivan, M.B., Martiny, A.C., Steglich, C., Barry, K., DeLong, E.F., Chisholm, S.W., 2006. Genomic islands and the ecology and evolution of *Prochlorococcus*. Science 311 (5768), 1768–1770.

Collins, J.R., Edwards, B.R., Thamatrakoln, K., Ossolinski, J.E., DiTullio, G.R., Bidle, K.D., Doney, S.C., Van Mooy, B.A., 2015. The multiple fates of sinking particles in the North Atlantic Ocean. Glob. Biogeochem. Cycles 29 (9), 1471–1494.

Cottrell, M.T., Suttle, C.A., 1995a. Genetic diversity of algal viruses which lyse the photosynthetic picoflagellate *Micromonas pusilla* (Prasinophyceae). Appl. Environ. Microbiol. 61 (8), 3088–3091.

Cottrell, M.T., Suttle, C.A., 1995b. Dynamics of lytic virus infecting the photosynthetic marine picoflagellate *Micromonas pusilla*. Limnol. Oceanogr. 40 (4), 730–739.

Croll, Donald A., Marinovic, Baldo, Chavez, Francisco P., Black, Nancy, Ternullo, Richard, Tershey, Bernie R., et al., 2005. From wind to whales: trophic links in a coastal upwelling system. Marine Ecology Progress Series 289, 117–130.

De Paepe, M., Taddei, F., 2006. Viruses' life history: towards a mechanistic basis of a trade-off between survival and reproduction among phages. PLoS Biol. 4 (7), e193.

De Smet, J., Zimmermann, M., Kogadeeva, M., Ceyssens, P.J., Vermaelen, W., Blasdel, B., Bin Jang, H., Sauer, U., Lavigne, R., 2016. High coverage metabolomics analysis reveals phage-specific alterations to *Pseudomonas aeruginosa* physiology during infection. ISME J. 10 (8), 1823–1835.

DeLong, J.P., Al-Sammak, M.A., Al-Ameeli, Z.T., Dunigan, D.D., Edwards, K.F., Fuhrmann, J.J., Gleghorn, J.P., Li, H., Haramoto, K., Harrison, A.O., Marston, M.F., 2022. Towards an integrative view of virus phenotypes. Nat. Rev. Microbiol. 20 (2), 83–94.

Demory, D., Arsenieff, L., Simon, N., Six, C., Rigaut-Jalabert, F., Marie, D., Ge, P., Bigeard, E., Jacquet, S., Sciandra, A., Bernard, O., Rabouille, S., Baudoux, A.C., 2017. Temperature is a key factor in *Micromonas*–virus interactions. ISME J. 11 (3), 601–612.

Deng, L.I., Hayes, P.K., 2008. Evidence for cyanophages active against bloom-forming freshwater cyanobacteria. Freshw. Biol. 53 (6), 1240–1252.

Derelle, E., Yau, S., Moreau, H., Grimsley, N.H., 2018. Prasinovirus attack of *Ostreococcus* is furtive by day but savage by night. J. Virol. 92 (4), e01703–e01717.

Diaz, B.P., Knowles, B., Johns, C.T., Laber, C.P., Bondoc, K.G.V., Haramaty, L., Natale, F., Harvey, E.L., Kramer, S.J., Bolaños, L.M., Lowenstein, D.P., Fredricks, H.F., Graff, J., Westberry, T.K., Mojica, K.D.A., Haëntjens, N., Baetge, N., Gaube, P., Boss, E., Carson, C.A., Behrenfeld, M.J., Van Mooy, B.A.S., Bidle, K.D., et al., 2021. Seasonal mixed layer depth shapes phytoplankton physiology, viral production, and accumulation in the North Atlantic. Nat. Commun. 12 (1), 1–16.

Dillon, A., Parry, J.D., 2008. Characterization of temperate cyanophages active against freshwater phycocyanin-rich *Synechococcus* species. Freshw. Biol. 53 (6), 1253–1261.

Dolan, J., McKeon, K., 2005. The reliability of grazing rate estimates from dilution experiments: Have we over-estimated rates of organic carbon consumption by microzooplankton? Ocean Sci. 1 (1), 1–7.

Doney, S.C., Fabry, V.J., Feely, R.A., Kleypas, J.A., 2009. Ocean acidification: the other CO2 problem. Annu. Rev. Mar. Sci. 1, 169–192.

Du Toit, A., 2018. Carbon export into the deep ocean. Nat. Rev. Microbiol. 16 (5), 260–261.

Du, S., Qin, F., Zhang, Z., Tian, Z., Yang, M., Liu, X., Zhao, G., Xia, Q., Zhao, Y., 2021. Genomic diversity, life strategies and ecology of marine HTVC010P-type pelagiphages. Microb. Genomics 7 (7).

Edwards, M., Richardson, A.J., 2004. Impact of climate change on marine pelagic phenology and trophic mismatch. Nature 430 (7002), 881–884.

Edwards, K.F., Steward, G.F., 2018. Host traits drive viral life histories across phytoplankton viruses. Am. Nat. 191 (5), 566–581.

Edwards, C.A., Batchelder, H.P., Powell, T.M., 2000. Modeling microzooplankton and macrozooplankton dynamics within a coastal upwelling system. J. Plankton Res. 22 (9), 1619–1648.

Eggleston, E.M., Hewson, I., 2016. Abundance of two *Pelagibacter* ubique bacteriophage genotypes along a latitudinal transect in the North and South Atlantic Oceans. Front. Microbiol. 7, 1534.

Eich, C., Pont, S.B., Brussaard, C.P., 2021. Effects of UV radiation on the chlorophyte *Micromonas polaris* host–virus interactions and MpoV-45T virus infectivity. Microorganisms 9 (12), 2429.

Eilers, H., Pernthaler, J., Peplies, J., Glöckner, F.O., Gerdts, G., Amann, R., 2001. Isolation of novel pelagic bacteria from the German Bight and their seasonal contributions to surface picoplankton. Appl. Environ. Microbiol. 67 (11), 5134–5142.

Evans, G.T., Paranjape, M.A., 1992. Precision of estimates of phytoplankton growth and microzooplankton grazing when the functional response of grazers may be nonlinear. Mar. Ecol. Prog. Ser. 80 (2), 285–290.

Evans, C., Archer, S.D., Jacquet, S., Wilson, W.H., 2003. Direct estimates of the contribution of viral lysis and microzooplankton grazing to the decline of a *Micromonas* spp. population. Aquat. Microb. Ecol. 30 (3), 207–219.

Farmer, J.R., Sigman, D.M., Granger, J., Underwood, O.M., Fripiat, F., Cronin, T.M., Martínez-García, A., Haug, G.H., 2021. Arctic Ocean stratification set by sea level and freshwater inputs since the last ice age. Nat. Geosci. 14 (9), 684–689.

Feely, R.A., Orr, J., Fabry, V.J., Kleypas, J.A., Sabine, C.L., Langdon, C., 2009. Present and future changes in seawater chemistry due to ocean acidification. In: Geophysical Monograph Series. American Geophysical Union, pp. 175–188.

Fenchel, T., 2008. The microbial loop–25 years later. J. Exp. Mar. Biol. Ecol. 366 (1-2), 99–103.

Field, C.B., Behrenfeld, M.J., Randerson, J.T., Falkowski, P., 1998. Primary production of the biosphere: integrating terrestrial and oceanic components. Science 281 (5374), 237–240.

Finkel, Z.V., Beardall, J., Flynn, K.J., Quigg, A., Rees, T.A.V., Raven, J.A., 2010. Phytoplankton in a changing world: cell size and elemental stoichiometry. J. Plankton Res. 32 (1), 119–137.

Flierl, G., Grünbaum, D., Levins, S., Olson, D., 1999. From individuals to aggregations: the interplay between behavior and physics. J. Theor. Biol. 196 (4), 397–454.

Flombaum, P., Gallegos, J.L., Gordillo, R.A., Rincón, J., Zabala, L.L., Jiao, N., Karl, D.M., Li, W.K.W., Lomas, M.W., Veneziano, D., Vera, C.S., Vrugt, J.A., Martiny, A.C., 2013. Present and future global distributions of the marine Cyanobacteria *Prochlorococcus* and *Synechococcus*. Proc. Natl. Acad. Sci. 110 (24), 9824–9829.

Flores, C.O., Meyer, J.R., Valverde, S., Farr, L., Weitz, J.S., 2011. Statistical structure of host–phage interactions. Proc. Natl. Acad. Sci. 108 (28), E288–E297.

Flores, C.O., Valverde, S., Weitz, J.S., 2013. Multi-scale structure and geographic drivers of cross-infection within marine bacteria and phages. ISME J. 7 (3), 520–532.

Flores-Uribe, J., Philosof, A., Sharon, I., Fridman, S., Larom, S., Béjà, O., 2019. A novel uncultured marine cyanophage lineage with lysogenic potential linked to a putative marine *Synechococcus* 'relic'prophage. Environ. Microbiol. Rep. 11 (4), 598–604.

Follows, M.J., Dutkiewicz, S., 2011. Modeling diverse communities of marine microbes. Annu. Rev. Mar. Sci. 3 (1), 427–451.

Follows, M.J., Dutkiewicz, S., Grant, S., Chisholm, S.W., 2007. Emergent biogeography of microbial communities in a model ocean. Science 315 (5820), 1843–1846.

Forcone, K., Coutinho, F.H., Cavalcanti, G.S., Silveira, C.B., 2021. Prophage genomics and ecology in the family Rhodobacteraceae. Microorganisms 9 (6), 1115.

Forterre, P., 2011. Manipulation of cellular syntheses and the nature of viruses: the virocell concept. Comptes Rendus Chimie 14 (4), 392–399.

Foulon, E., Not, F., Jalabert, F., Cariou, T., Massana, R., Simon, N., 2008. Ecological niche partitioning in the picoplanktonic green alga *Micromonas pusilla*: evidence from environmental surveys using phylogenetic probes. Environ. Microbiol. 10 (9), 2433–2443.

Fox, J., Behrenfeld, M.J., Haëntjens, N., Chase, A., Kramer, S.J., Boss, E., Karp-Boss, L., Fisher, N.L., Penta, W.B., Westberry, T.K., Halsey, K.H., 2020. Phytoplankton growth and productivity in the Western North Atlantic: Observations of regional variability from the NAAMES field campaigns. Front. Mar. Sci. 7, 24.

Frada, M.J., Schatz, D., Farstey, V., Ossolinski, J.E., Sabanay, H., Ben-Dor, S., Koren, I., Vardi, A., 2014. Zooplankton may serve as transmission vectors for viruses infecting algal blooms in the ocean. Curr. Biol. 24 (21), 2592–2597.

Franche, C., 1987. Isolation and characterization of a temperate cyanophage for a tropical Anabaena strain. Arch. Microbiol. 148 (3), 172–177.

Franks, P.J., Inman, B.G., Mac Kinnon, J.A., Alford, M.H., Waterhouse, A.F., 2022. Oceanic turbulence from a planktonic perspective. Limnol. Oceanogr. 67 (2), 348–363.

Fridman, S., Flores-Uribe, J., Larom, S., Alalouf, O., Liran, O., Yacoby, I., Salama, F., Bailleul, B., Rappaport, F., Ziv, T., Sharon, I., 2017. A myovirus encoding both photosystem I and II proteins enhances cyclic electron flow in infected *Prochlorococcus* cells. Nat. Microbiol. 2 (10), 1350–1357.

Fuchs, H.L., Gerbi, G.P., 2016. Seascape-level variation in turbulence-and wave-generated hydrodynamic signals experienced by plankton. Prog. Oceanogr. 141, 109–129.

Fuchsman, C.A., Palevsky, H.I., Widner, B., Duffy, M., Carlson, M.C., Neibauer, J.A., Mulholland, M.R., Keil, R.G., Devol, A.H., Rocap, G., 2019. Cyanobacteria and cyanophage contributions to carbon and nitrogen cycling in an oligotrophic oxygen-deficient zone. ISME J. 13 (11), 2714–2726.

Fuchsman, C.A., Carlson, M.C., Garcia Prieto, D., Hays, M.D., Rocap, G., 2021. Cyanophage host-derived genes reflect contrasting selective pressures with depth in the oxic and anoxic water column of the Eastern Tropical North Pacific. Environ. Microbiol. 23 (6), 2782–2800.

Fuhrman, J.A., 1999. Marine viruses and their biogeochemical and ecological effects. Nature 399 (6736), 541–548.

Fuhrman, J.A., Lee, S.H., Masuchi, Y., Davis, A.A., Wilcox, R.M., 1994. Characterization of marine prokaryotic communities via DNA and RNA. Microb. Ecol. 28 (2), 133–145.

Fuhrmann, M., Richard, G., Quere, C., Petton, B., Pernet, F., 2019. Low pH reduced survival of the oyster *Crassostrea gigas* exposed to the Ostreid herpesvirus 1 by altering the metabolic response of the host. Aquaculture 503, 167–174.

Fulton, J.M., Fredricks, H.F., Bidle, K.D., Vardi, A., Kendrick, B.J., DiTullio, G.R., Van Mooy, B.A., 2014. Novel molecular determinants of viral susceptibility and resistance in the lipidome of *Emiliania huxleyi*. Environ. Microbiol. 16 (4), 1137–1149.

Gao, E.B., Yuan, X.P., Li, R.H., Zhang, Q.Y., 2009. Isolation of a novel cyanophage infectious to the filamentous cyanobacterium *Planktothrix agardhii* (Cyanophyceae) from Lake Donghu, China. Aquat. Microb. Ecol. 54 (2), 163–170.

Gao, E.B., Huang, Y., Ning, D., 2016. Metabolic genes within cyanophage genomes: implications for diversity and evolution. Gene 7 (10), 80.

Gardiner, C.W., 2009. Handbook of Stochastic Methods for Physics, Chemistry, and the Natural Sciences, fourth ed. Springer, Berlin.

Garza, D.R., Suttle, C.A., 1998. The effect of cyanophages on the mortality of *Synechococcus* spp. and selection for UV resistant viral communities. Microb. Ecol. 36 (3), 281–292.

Gasper, R., Schwach, J., Hartmann, J., Holtkamp, A., Wiethaus, J., Riedel, N., Hofmann, E., Frankenberg-Dinkel, N., 2017. Distinct features of cyanophage-encoded T-type phycobiliprotein lyase ΦCpeT: the role of auxiliary metabolic genes. J. Biol. Chem. 292 (8), 3089–3098.

Geng, H., Belas, R., 2010. Molecular mechanisms underlying Roseobacter–phytoplankton symbioses. Curr. Opin. Biotechnol. 21 (3), 332–338.

Giovannoni, S.J., 2017. SAR11 bacteria: the most abundant plankton in the oceans. Annu. Rev. Mar. Sci. 9, 231–255.

Giovannoni, S.J., DeLong, E.F., Schmidt, T.M., Pace, N.R., 1990. Tangential flow filtration and preliminary phylogenetic analysis of marine picoplankton. Appl. Environ. Microbiol. 56 (8), 2572–2575.

Gledhill, M., Achterberg, E.P., Li, K., Mohamed, K.N., Rijkenberg, M.J., 2015. Influence of ocean acidification on the complexation of iron and copper by organic ligands in estuarine waters. Mar. Chem. 177, 421–433.

Glibert, P.M., Icarus Allen, J., Artioli, Y., Beusen, A., Bouwman, L., Harle, J., Holmes, R., Holt, J., 2014. Vulnerability of coastal ecosystems to changes in harmful algal bloom distribution in response to climate change: projections based on model analysis. Glob. Chang. Biol. 20 (12), 3845–3858.

Gobler, C.J., 2020. Climate change and harmful algal blooms: insights and perspective. Harmful Algae 91, 101731.

Gobler, C.J., Hutchins, D.A., Fisher, N.S., Cosper, E.M., Sañudo-Wilhelmy, S.A., 1997. Release and bioavailability of C, N, P Se, and Fe following viral lysis of a marine chrysophyte. Limnol. Oceanogr. 42 (7), 1492–1504.

Gobler, C.J., Doherty, O.M., Hattenrath-Lehmann, T.K., Griffith, A.W., Kang, Y., Litaker, R.W., 2017. Ocean warming since 1982 has expanded the niche of toxic algal blooms in the North Atlantic and North Pacific oceans. Proc. Natl. Acad. Sci. 114 (19), 4975–4980.

Golec, P., Karczewska-Golec, J., Łoś, M., Węgrzyn, G., 2014. Bacteriophage T4 can produce progeny virions in extremely slowly growing *Escherichia coli* host: comparison of a mathematical model with the experimental data. FEMS Microbiol. Lett. 351 (2), 156–161.

Graff, J.R., Behrenfeld, M.J., 2018. Photoacclimation responses in subarctic Atlantic phytoplankton following a natural mixing-restratification event. Front. Mar. Sci. 5, 209.

Grasso, C.R., Pokrzywinski, K.L., Waechter, C., Rycroft, T., Zhang, Y., Aligata, A., Kramer, M., Lamsal, A., 2022. A review of cyanophage–host relationships: highlighting cyanophages as a potential cyanobacteria control strategy. Toxins 14 (6), 385.

Gregory, A.C., Zayed, A.A., Conceição-Neto, N., Temperton, B., Bolduc, B., Alberti, A., Ardyna, M., Arkhipova, K., Carmichael, M., Cruaud, C., Dimier, C., Domínguez-Huerta, G., Ferland, J., Kandels, S., Liu, Y., Marec, C., Pesant, S., Picheral, M., Pisarev, S., Poulain, J., Tremblay, J.-É., Vik, D., Coordinators, T.O., Babin, M., Bowler, C., Culley, A.I., de Vargas, C., Dutilh, B.E., Iudicone, D., Karp-Boss, L., Roux, S., Sunagawa, S., Wincker, P., Sullivan, M.B., 2019. Marine DNA viral macro-and microdiversity from pole to pole. Cell 177 (5), 1109–1123.

Gruber, N., Clement, D., Carter, B.R., Feely, R.A., van Heuven, S., Hoppema, M., Ishii, M., Key, R.M., Kozyr, A., Lauvset, S.K., Lo Monaco, C., Mathis, J.T., Murata, A., Olsen, A., Perez, F.F., Sabine, C.L., Tanhua, T., Wanninkhof, R., 2019. The oceanic sink for anthropogenic CO2 from 1994 to 2007. Science 363 (6432), 1193–1199.

Grzymski, J.J., Dussaq, A.M., 2012. The significance of nitrogen cost minimization in proteomes of marine microorganisms. ISME J. 6 (1), 71–80.

Guidi, L., Chaffron, S., Bittner, L., Eveillard, D., Larhlimi, A., Roux, S., Darzi, Y., Audic, S., Berline, L., Brum, J.R., Coelho, L.P., Espinoza, J.C.I., Malviya, S., Sunagawa, S., Dimier, C., Kandels-Lewis, S., Picheral, M., Poulain, J., Searson, S., Stemmann, L., Not, F., Hingamp, P., Speich, S., Follows, M., Karp-Boss, L., Boss, E., Ogata, H., Pesant, S., Weissenbach, J., Wincker, P., Acinas, S.G., Bork, P., de Vargas, C., Iudicone, D., Sullivan, M.B., Raes, J., Karsenti, E., Bowler, C., Gorsky, G., Tara Oceans Consortium Coordinators, 2016. Plankton networks driving carbon export in the oligotrophic ocean. Nature 532 (7600), 465–470.

Guillou, L., Eikrem, W., Chrétiennot-Dinet, M.J., Le Gall, F., Massana, R., Romari, K., Pedrós-Alió, C., Vaulot, D., 2004. Diversity of picoplanktonic prasinophytes assessed by direct nuclear SSU rDNA sequencing of environmental samples and novel isolates retrieved from oceanic and coastal marine ecosystems. Protist 155 (2), 193–214.

Guinotte, J.M., Fabry, V.J., 2008. Ocean acidification and its potential effects on marine ecosystems. Ann. N. Y. Acad. Sci. 1134 (1), 320–342.

Haaber, J., Middelboe, M., 2009. Viral lysis of *Phaeocystis pouchetii*: implications for algal population dynamics and heterotrophic C, N and P cycling. ISME J. 3 (4), 430–441.

Hadas, H., Einav, M., Fishov, I., Zaritsky, A., 1997. Bacteriophage T4 development depends on the physiology of its host *Escherichia coli*. Microbiology 143 (1), 179–185.

Harrison, A.O., Moore, R.M., Polson, S.W., Wommack, K.E., 2019. Reannotation of the ribonucleotide reductase in a cyanophage reveals life history strategies within the virioplankton. Front. Microbiol. 10, 134.

Hasumi, H., Nagata, T., 2014. Modeling the global cycle of marine dissolved organic matter and its influence on marine productivity. Ecol. Model. 288, 9–24.

Heiskanen, A.S., 1993. Mass encystment and sinking of dinoflagellates during a spring bloom. Mar. Biol. 116 (1), 161–167.

Hewson, I., Govil, S.R., Capone, D.G., Carpenter, E.J., Fuhrman, J.A., 2004. Evidence of *Trichodesmium* viral lysis and potential significance for biogeochemical cycling in the oligotrophic ocean. Aquat. Microb. Ecol. 36 (1), 1–8.

Howard-Varona, C., Lindback, M.M., Bastien, G.E., Solonenko, N., Zayed, A.A., Jang, H., Andreopoulos, B., Brewer, H.M., Glavina del Rio, T., Adkins, J.N., Paul, S., 2020. Phage-specific metabolic reprogramming of virocells. ISME J. 14 (4), 881–895.

Hu, N.T., Thiel, T., Giddings Jr., T.H., Wolk, C.P., 1981. New Anabaena and Nostoc cyanophages from sewage settling ponds. Virology 114 (1), 236–246.

Huang, S., Wang, K., Jiao, N., Chen, F., 2012. Genome sequences of siphoviruses infecting marine *Synechococcus* unveil a diverse cyanophage group and extensive phage–host genetic exchanges. Environ. Microbiol. 14 (2), 540–558.

Huang, X., Jiao, N., Zhang, R., 2021. The genomic content and context of auxiliary metabolic genes in roseophages. Environ. Microbiol. 23 (7), 3743–3757.

Huey, R.B., Slatkin, M., 1976. Cost and benefits of lizard thermoregulation. Q. Rev. Biol. 51 (3), 363–384.

Hughes, B.S., Cullum, A.J., Bennett, A.F., 2007. An experimental evolutionary study on adaptation to temporally fluctuating pH in *Escherichia coli*. Physiol. Biochem. Zool. 80 (4), 406–421.

Hunter, J.E., Frada, M.J., Fredricks, H.F., Vardi, A., Van Mooy, B.A., 2015. Targeted and untargeted lipidomics of *Emiliania huxleyi* viral infection and life cycle phases highlights molecular biomarkers of infection, susceptibility, and ploidy. Front. Mar. Sci. 2, 81.

Hurwitz, B.L., Sullivan, M.B., 2013. The Pacific Ocean Virome (POV): a marine viral metagenomic dataset and associated protein clusters for quantitative viral ecology. PLoS One 8 (2), e57355.

Hurwitz, B.L., U'Ren, J.M., 2016. Viral metabolic reprogramming in marine ecosystems. Curr. Opin. Microbiol. 31, 161–168.

Hurwitz, B.L., Westveld, A.H., Brum, J.R., Sullivan, M.B., 2014. Modeling ecological drivers in marine viral communities using comparative metagenomics and network analyses. Proc. Natl. Acad. Sci. 111 (29), 10714–10719.

Jacobsen, A., Bratbak, G., Heldal, M., 1996. Isolation and characterization of a virus infecting *phaeocystis pouchetii* (prymnesiophyceae) 1. J. Phycol. 32 (6), 923–927.

Jasti, S., Sieracki, M.E., Poulton, N.J., Giewat, M.W., Rooney-Varga, J.N., 2005. Phylogenetic diversity and specificity of bacteria closely associated with Alexandrium spp. and other phytoplankton. Appl. Environ. Microbiol. 71 (7), 3483–3494.

Ji, P., Zhang, Y., Wang, J., Zhao, F., 2017. Meta Sort untangles metagenome assembly by reducing microbial community complexity. Nat. Commun. 8 (1), 1–14.

Jiao, N., Herndl, G.J., Hansell, D.A., Benner, R., Kattner, G., Wilhelm, S.W., Kirchman, D.L., Weinbauer, M.G., Luo, T., Chen, F., Azam, F., 2010. Microbial production of recalcitrant dissolved organic matter: long-term carbon storage in the global ocean. Nat. Rev. Microbiol. 8 (8), 593–599.

John, S.G., Mendez, C.B., Deng, L., Poulos, B., Kauffman, A.K.M., Kern, S., Brum, J., Polz, M.F., Boyle, E.A., Sullivan, M.B., 2011. A simple and efficient method for concentration of ocean viruses by chemical flocculation. Environ. Microbiol. Rep. 3 (2), 195–202.

Johns, C.T., Grubb, A.R., Nissimov, J.I., Natale, F., Knapp, V., Mui, A., Fredricks, H.F., Van Mooy, B.A., Bidle, K.D., 2019. The mutual interplay between calcification and coccolithovirus infection. Environ. Microbiol. 21 (6), 1896–1915.

Jonkers, H.M., Abed, R.M., 2003. Identification of aerobic heterotrophic bacteria from the photic zone of a hypersaline microbial mat. Aquat. Microb. Ecol. 30 (2), 127–133.

Jungblut, A.D., Raymond, F., Dion, M.B., Moineau, S., Mohit, V., Nguyen, G.Q., Déraspe, M., Francovic-Fontaine, É., Lovejoy, C., Culley, A.I., Corbeil, J., 2021. Genomic diversity and CRISPR-Cas systems in the cyanobacterium Nostoc in the High Arctic. Environ. Microbiol. 23 (6), 2955–2968.

Juranek, L.W., White, A.E., Dugenne, M., Henderikx Freitas, F., Dutkiewicz, S., Ribalet, F., Ferrón, S., Armbrust, E.V., Karl, D.M., 2020. The importance of the phytoplankton "middle class" to ocean net community production. Glob. Biogeochem. Cycles 34 (12), e2020GB006702.

Kana, T.M., Gilbert, P.M., 1987. Effect of irradiances up to 2000 μE m^{-2} s^{-1} on marine *Synechococcus* WH7803—I. Growth, pigmentation, and cell composition. Deep sea research Part A. Oceanogr. Res. Papers 34 (4), 479–495.

Kana, T.M., Glibert, P.M., Goericke, R., Welschmeyer, N.A., 1988. Zeaxanthin and ß-carotene in *Synechococcus* WH7803 respond differently to irradiance. Limnol. Oceanogr. 33 (6 Pt. 2), 1623–1626.

Kaneko, H., Blanc-Mathieu, R., Endo, H., Chaffron, S., Delmont, T.O., Gaia, M., Henry, N., Hernández-Velázquez, R., Nguyen, C.H., Mamitsuka, H., Forterre, P., Jaillon, O., de Vargas, C., Sullivan, M.B., Suttle, C.A., Guidi, L., Ogata, H., 2021. Eukaryotic virus composition can predict the efficiency of carbon export in the global ocean. Iscience 24 (1), 102002.

Kang, I., Oh, H.M., Kang, D., Cho, J.C., 2013. Genome of a SAR116 bacteriophage shows the prevalence of this phage type in the oceans. Proc. Natl. Acad. Sci. 110 (30), 12343–12348.

Karl, D.M., 1999. A sea of change: biogeochemical variability in the North Pacific Subtropical Gyre. Ecosystems 2 (3), 181–214.

Kassen, R., 2002. The experimental evolution of specialists, generalists, and the maintenance of diversity. J. Evol. Biol. 15 (2), 173–190.

Kathuria, S., Martiny, A.C., 2011. Prevalence of a calcium-based alkaline phosphatase associated with the marine cyanobacterium *Prochlorococcus* and other ocean bacteria. Environ. Microbiol. 13 (1), 74–83.

Kauffman, K.M., Chang, W.K., Brown, J.M., Hussain, F.A., Yang, J., Polz, M.F., Kelly, L., 2022. Resolving the structure of phage–bacteria interactions in the context of natural diversity. Nat. Commun. 13 (1), 1–20.

Keller, D.P., Hood, R.R., 2011. Modeling the seasonal autochthonous sources of dissolved organic carbon and nitrogen in the upper Chesapeake Bay. Ecol. Model. 222 (5), 1139–1162.

Keller, D.P., Hood, R.R., 2013. Comparative simulations of dissolved organic matter cycling in idealized oceanic, coastal, and estuarine surface waters. J. Mar. Syst. 109, 109–128.

Kelly, L., Ding, H., Huang, K.H., Osburne, M.S., Chisholm, S.W., 2013. Genetic diversity in cultured and wild marine cyanomyoviruses reveals phosphorus stress as a strong selective agent. ISME J. 7 (9), 1827–1841.

Kendrick, B.J., DiTullio, G.R., Cyronak, T.J., Fulton, J.M., Van Mooy, B.A., Bidle, K.D., 2014. Temperature-induced viral resistance in *Emiliania huxleyi* (Prymnesiophyceae). PLoS One 9 (11), e112134.

Keown, R.A., Dums, J.T., Brumm, P.J., MacDonald, J., Mead, D.A., Ferrell, B.D., Moore, R.M., Harrison, A.O., Polson, S.W., Wommack, K.E., 2022. Novel viral DNA polymerases from metagenomes suggest genomic sources of strand-displacing biochemical phenotypes. Front. Microbiol. 13, 858366.

Kimura, K., Tomaru, Y., 2015. Discovery of two novel viruses expands the diversity of single-stranded DNA and single-stranded RNA viruses infecting a cosmopolitan marine diatom. Appl. Environ. Microbiol. 81 (3), 1120–1131.

Kiørboe, T., Saiz, E., 1995. Planktivorous feeding in calm and turbulent environments, with emphasis on copepods. Mar. Ecol. Prog. Ser. *122*, 135–145.

Klaas, C., Archer, D.E., 2002. Association of sinking organic matter with various types of mineral ballast in the deep sea: implications for the rain ratio. Glob. Biogeochem. Cycles 16 (4), 63.

Knowles, B., Bonachela, J.A., Behrenfeld, M.J., Bondoc, K.G., Cael, B.B., Carlson, C.A., Cieslik, N., Diaz, B., Fuchs, H.L., Graff, J.R., Grasis, J.A., Halsey, K.H., Haramaty, L., Johns, C.T., Natale, F., Nissimov, J.I., Schieler, B., Thamatrakoln, K., Frede Thingstad, T., Våge, S., Watkins, C., Westberry, T.K., Bidle, K.D., 2020. Temperate infection in a virus–host system previously known for virulent dynamics. Nat. Commun. 11 (1), 1–13.

Kranzler, C.F., Krause, J.W., Brzezinski, M.A., Edwards, B.R., Biggs, W.P., Maniscalco, M., McCrow, J.P., Van Mooy, B.A., Bidle, K.D., Allen, A.E., Thamatrakoln, K., 2019. Silicon limitation facilitates virus infection and mortality of marine diatoms. Nat. Microbiol. 4 (11), 1790–1797.

Kranzler, C.F., Brzezinski, M.A., Cohen, N.R., Lampe, R.H., Maniscalco, M., Till, C.P., Mack, J., Latham, J.R., Bruland, K.W., Twining, B.S., Marchetti, A., 2021. Impaired viral infection and reduced mortality of diatoms in iron-limited oceanic regions. Nat. Geosci. 14 (4), 231–237.

Kroeker, K.J., Kordas, R.L., Crim, R., Hendriks, I.E., Ramajo, L., Singh, G.S., Duarte, C.M., Gattuso, J.P., 2013. Impacts of ocean acidification on marine organisms: quantifying sensitivities and interaction with warming. Glob. Chang. Biol. 19 (6), 1884–1896.

Kuznecova, J., Šulčius, S., Vogts, A., Voss, M., Jürgens, K., Šimoliūnas, E., 2020. Nitrogen flow in diazotrophic cyanobacterium Aphanizomenon flos-aquae is altered by cyanophage infection. Front. Microbiol. 11, 2010.

Laber, C.P., Hunter, J.E., Carvalho, F., Collins, J.R., Hunter, E.J., Schieler, B.M., Boss, E., More, K., Frada, M., Thamatrakoln, K., Brown, C.M., Haramaty, L., Ossolinski, J., Fredricks, H., Nissimov, J.I., Vandzura, R., Sheyn, U., Lehahn, Y., Chant, R.J., Martins, A.M., Coolen, M.J.L., Vardi, A., DiTullio, G.R., Van Mooy, B.A.S., Bidle, K.D., 2018. Coccolithovirus facilitation of carbon export in the North Atlantic. Nat. Microbiol. 3 (5), 537–547.

Labrenz, M., Collins, M.D., Lawson, P.A., Tindall, B.J., Braker, G., Hirsch, P., 1998. *Antarctobacter heliothermus* gen. nov., sp. nov., a budding bacterium from hypersaline and heliothermal Ekho Lake. Int. J. Syst. Evol. Microbiol. 48 (4), 1363–1372.

Landry, M.R., Hassett, R.P.L., 1982. Estimating the grazing impact of marine micro-zooplankton. Mar. Biol. 67 (3), 283–288.

Larsen, A., Castberg, T., Sandaa, R.A., Brussaard, C.P.D., Egge, J., Heldal, M., Paulino, A., Thyrhaug, R., Van Hannen, E.J., Bratbak, G., 2001. Population dynamics and diversity of phytoplankton, bacteria and viruses in a seawater enclosure. Mar. Ecol. Prog. Ser. 221, 47–57.

Lau, S.C., Tsoi, M.M., Li, X., Plakhotnikova, I., Wu, M., Wong, P.K., Qian, P.Y., 2004. *Loktanella hongkongensis* sp. nov., a novel member of the α-Proteobacteria originating from marine biofilms in Hong Kong waters. Int. J. Syst. Evol. Microbiol. 54 (6), 2281–2284.

Le Quéré, C., Raupach, M.R., Canadell, J.G., Marland, G., Bopp, L., Ciais, P., Conway, T.J., Doney, S.C., Feely, R.A., Orr, J., Fabry, V.J., Kleypas, J.A., Sabine, C.L., Langdon, C., 2009. Present and future changes in seawater chemistry due to ocean acidification. In: Geophysical Monograph Series. American Geophysical Union, pp. 175–188.

Lee, L.H., Lui, D., Platner, P.J., Hsu, S.F., Chu, T.C., Gaynor, J.J., Vega, Q.C., Lustigman, B.K., 2006. Induction of temperate cyanophage AS-1 by heavy metal–copper. BMC Microbiol. 6 (1), 1–7.

Lehahn, Y., Koren, I., Schatz, D., Frada, M., Sheyn, U., Boss, E., Efrati, S., Rudich, Y., Trainic, M., Sharoni, S., Laber, C., 2014. Decoupling physical from biological processes to assess the impact of viruses on a mesoscale algal bloom. Curr. Biol. 24 (17), 2041–2046.

Levin, S.A., 1992. The problem of pattern and scale in ecology: the Robert H. MacArthur award lecture. Ecology 73 (6), 1943–1967.

Levins, R., 1968. Evolution in Changing Environments: Some Theoretical Exploration. Princeton University Press, Princeton, New Jersey.

Lewis, L.A., McCourt, R.M., 2004. Green algae and the origin of land plants. Am. J. Bot. 91 (10), 1535–1556.

Li, D., Luo, R., Liu, C.M., Leung, C.M., Ting, H.F., Sadakane, K., Yamashita, H., Lam, T.W., 2016a. MEGAHIT v1. 0: a fast and scalable metagenome assembler driven by advanced methodologies and community practices. Methods 102, 3–11.

Li, B., Zhang, S., Long, L., Huang, S., 2016b. Characterization and complete genome sequences of three N4-like roseobacter phages isolated from the South China Sea. Curr. Microbiol. 73 (3), 409–418.

Li, G., Cheng, L., Zhu, J., Trenberth, K.E., Mann, M.E., Abraham, J.P., 2020. Increasing ocean stratification over the past half-century. Nat. Clim. Chang. 10 (12), 1116–1123.

Liang, K.C., Sakakibara, Y., 2021. MetaVelvet-DL: a MetaVelvet deep learning extension for de novo metagenome assembly. BMC bioinformatics 22 (6), 1–21.

Lindell, D., Jaffe, J.D., Johnson, Z.I., Church, G.M., Chisholm, S.W., 2005. Photosynthesis genes in marine viruses yield proteins during host infection. Nature 438 (7064), 86–89.

Lindh, M.V., Riemann, L., Baltar, F., Romero-Oliva, C., Salomon, P.S., Granéli, E., Pinhassi, J., 2013. Consequences of increased temperature and acidification on bacterioplankton community composition during a mesocosm spring bloom in the Baltic Sea. Environ. Microbiol. Rep. 5 (2), 252–262.

Litchman, E., Klausmeier, C.A., 2008. Trait-based community ecology of phytoplankton. Annu. Rev. Ecol. Evol. Syst., 615–639.

Lønborg, C., Middelboe, M., Brussaard, C.P., 2013. Viral lysis of *Micromonas pusilla*: impacts on dissolved organic matter production and composition. Biogeochemistry 116 (1), 231–240.

Ma, X., Coleman, M.L., Waldbauer, J.R., 2018. Distinct molecular signatures in dissolved organic matter produced by viral lysis of marine cyanobacteria. Environ. Microbiol. 20 (8), 3001–3011.

Maat, D.S., Brussaard, C.P., 2016. Both phosphorus and nitrogen limitation constrain viral proliferation in marine phytoplankton. Aquat. Microb. Ecol. 77 (2), 87–97.

Maat, D.S., Crawfurd, K.J., Timmermans, K.R., Brussaard, C.P., 2014. Elevated CO2 and phosphate limitation favor *Micromonas pusilla* through stimulated growth and reduced viral impact. Appl. Environ. Microbiol. 80 (10), 3119–3127.

Maat, D.S., van Bleijswijk, J.D., Witte, H.J., Brussaard, C.P., 2016. Virus production in phosphorus-limited *Micromonas pusilla* stimulated by a supply of naturally low concentrations of different phosphorus sources, far into the lytic cycle. FEMS Microbiol. Ecol. 92 (9).

Maat, D.S., Biggs, T., Evans, C., Van Bleijswijk, J.D., Van der Wel, N.N., Dutilh, B.E., Brussaard, C.P., 2017. Characterization and temperature dependence of Arctic *Micromonas polaris* viruses. Viruses 9 (6), 134.

Mackinder, L.C., Worthy, C.A., Biggi, G., Hall, M., Ryan, K.P., Varsani, A., Harper, G.M., Wilson, W.H., Brownlee, C., Schroeder, D.C., 2009. A unicellular algal virus, *Emiliania huxleyi* virus 86, exploits an animal-like infection strategy. J. Gen. Virol. 90 (9), 2306–2316.

MacLeod, R.A., 1965. The question of the existence of specific marine bacteria. Bacteriol. Rev. 29 (1), 9–23.

Maidanik, I., Kirzner, S., Pekarski, I., Arsenieff, L., Tahan, R., Carlson, M.C., Shitrit, D., Baran, N., Goldin, S., Weitz, J.S., Lindell, D., 2022. Cyanophages from a less virulent clade dominate over their sister clade in global oceans. ISME J. 1–12.

Malmstrom, R.R., Kiene, R.P., Cottrell, M.T., Kirchman, D.L., 2004. Contribution of SAR11 bacteria to dissolved dimethylsulfoniopropionate and amino acid uptake in the North Atlantic Ocean. Appl. Environ. Microbiol. 70 (7), 4129–4135.
Malmstrom, R.R., Cottrell, M.T., Elifantz, H., Kirchman, D.L., 2005. Biomass production and assimilation of dissolved organic matter by SAR11 bacteria in the Northwest Atlantic Ocean. Appl. Environ. Microbiol. 71 (6), 2979–2986.
Malmstrom, R.R., Rodrigue, S., Huang, K.H., Kelly, L., Kern, S.E., Thompson, A., Roggensack, S., Berube, P.M., Henn, M.R., Chisholm, S.W., 2013. Ecology of uncultured *Prochlorococcus* clades revealed through single-cell genomics and biogeographic analysis. ISME J. 7 (1), 184–198.
Malviya, S., Scalco, E., Audic, S., Vincent, F., Veluchamy, A., Poulain, J., Wincker, P., Iudicone, D., de Vargas, C., Bittner, L., Zingone, A., 2016. Insights into global diatom distribution and diversity in the world's ocean. Proc. Natl. Acad. Sci. 113 (11), E1516–E1525.
Mankiewicz-Boczek, J., Jaskulska, A., Pawełczyk, J., Gągała, I., Serwecińska, L., Dziadek, J., 2016. Cyanophages infection of Microcystis bloom in lowland dam reservoir of Sulejów, Poland. Microb. Ecol. 71 (2), 315–325.
Mann, N.H., Clokie, M.R., 2012. Cyanophages. In: Ecology of Cyanobacteria II. Springer, Dordrecht, pp. 535–557.
Mann, N.H., Cook, A., Millard, A., Bailey, S., Clokie, M., 2003. Bacterial photosynthesis genes in a virus. Nature 424 (6950), 741–742.
Marei, E.M., Elbaz, R.M., Hammad, A.M.M., 2013. Induction of temperate cyanophages using heavy metal-copper. Int. J. Microbiol. Res. 5 (5), 472.
Margalef, R., 1998. Turbulence and marine life. Oceanogr. Lit. Rev. 3 (45), 499.
Marine, R., McCarren, C., Vorrasane, V., Nasko, D., Crowgey, E., Polson, S.W., Wommack, K.E., 2014. Caught in the middle with multiple displacement amplification: the myth of pooling for avoiding multiple displacement amplification bias in a metagenome. Microbiome 2 (1), 1–8.
Marston, M.F., Sallee, J.L., 2003. Genetic diversity and temporal variation in the cyanophage community infecting marine *Synechococcus* species in Rhode Island's coastal waters. Appl. Environ. Microbiol. 69 (8), 4639–4647.
Martinez-Hernandez, F., Fornas, O., Lluesma Gomez, M., Bolduc, B., de La Cruz Peña, M.J., Martínez, J.M., Anton, J., Gasol, J.M., Rosselli, R., Rodriguez-Valera, F., Sullivan, M.B., Acinas, S.G., Martinez-Garcia, M., 2017. Single-virus genomics reveals hidden cosmopolitan and abundant viruses. Nat. Commun. 8 (1), 1–13.
Martinez-Hernandez, F., Garcia-Heredia, I., Lluesma Gomez, M., Maestre-Carballa, L., Martínez Martínez, J., Martinez-Garcia, M., 2019. Droplet digital PCR for estimating absolute abundances of widespread Pelagibacter viruses. Front. Microbiol. 10, 1226.
Mateus, M.D., 2017. Bridging the gap between knowing and modeling viruses in marine systems—an upcoming frontier. Front. Mar. Sci. 3, 284.
Mayer, J.A., Taylor, F.J.R., 1979. A virus which lyses the marine nanoflagellate *Micromonas pusilla*. Nature 281 (5729), 299–301.
McDaniel, L.D., Paul, J.H., 2006. Temperate and lytic cyanophages from the Gulf of Mexico. J. Mar. Biol. Assoc. U. K. 86 (3), 517–527.
McDaniel, L., Houchin, L.A., Williamson, S.J., Paul, J.H., 2002. Lysogeny in marine *Synechococcus*. Nature 415 (6871), 496.
McMullen, A., Martinez-Hernandez, F., Martinez-Garcia, M., 2019. Absolute quantification of infecting viral particles by chip-based digital polymerase chain reaction. Environ. Microbiol. Rep. 11 (6), 855–860.
McNair, H.M., Brzezinski, M.A., Krause, J.W., 2018. Diatom populations in an upwelling environment decrease silica content to avoid growth limitation. Environ. Microbiol. 20 (11), 4184–4193.

Mészáros, L., Van der Meulen, F., Jongbloed, G., El Serafy, G., 2021. Climate change induced trends and uncertainties in phytoplankton spring bloom dynamics. Front. Mar. Sci. 8, 1067.
Middelboe, M., Lyck, P.G., 2002. Regeneration of dissolved organic matter by viral lysis in marine microbial communities. Aquat. Microb. Ecol. 27 (2), 187–194.
Middelboe, M., Jorgensen, N., Kroer, N., 1996. Effects of viruses on nutrient turnover and growth efficiency of noninfected marine bacterioplankton. Appl. Environ. Microbiol. 62 (6), 1991–1997.
Millard, A.D., Mann, N.H., 2006. A temporal and spatial investigation of cyanophage abundance in the Gulf of Aqaba, Red Sea. J. Mar. Biol. Assoc. U. K. 86 (3), 507–515.
Moebus, K., Nattkemper, H., 1981. Bacteriophage sensitivity patterns among bacteria isolated from marine waters. Helgoländer Meeresun. 34 (3), 375–385.
Morison, F., Harvey, E., Franzè, G., Menden-Deuer, S., 2019. Storm-induced predator-prey decoupling promotes springtime accumulation of North Atlantic phytoplankton. Front. Mar. Sci. 6, 608.
Morris, R.M., Rappé, M.S., Connon, S.A., Vergin, K.L., Siebold, W.A., Carlson, C.A., Giovannoni, S.J., 2002. SAR11 clade dominates ocean surface bacterioplankton communities. Nature 420 (6917), 806–810.
Morris, R.M., Cain, K.R., Hvorecny, K.L., Kollman, J.M., 2020. Lysogenic host–virus interactions in SAR11 marine bacteria. Nat. Microbiol. 5 (8), 1011–1015.
Mueller, J.A., Culley, A.I., Steward, G.F., 2014. Variables influencing extraction of nucleic acids from microbial plankton (viruses, bacteria, and protists) collected on nanoporous aluminum oxide filters. Appl. Environ. Microbiol. 80 (13), 3930–3942.
Murray, A.G., Jackson, G.A., 1992. Viral dynamics: a model of the effects of size, shape, motion and abundance of single-celled planktonic organisms and other particles. Mar. Ecol. Prog. Ser., 103–116.
Nagasaki, K., Tomaru, Y., Tarutani, K., Katanozaka, N., Yamanaka, S., Tanabe, H., Yamaguchi, M., 2003. Growth characteristics and intraspecies host specificity of a large virus infecting the dinoflagellate *Heterocapsa circularisquama*. Appl. Environ. Microbiol. 69 (5), 2580–2586.
Nagasaki, K., Tomaru, Y., Katanozaka, N., Shirai, Y., Nishida, K., Itakura, S., Yamaguchi, M., 2004. Isolation and characterization of a novel single-stranded RNA virus infecting the bloom-forming diatom *Rhizosolenia setigera*. Appl. Environ. Microbiol. 70 (2), 704–711.
Namiki, T., Hachiya, T., Tanaka, H., Sakakibara, Y., 2011. MetaVelvet: an extension of Velvet assembler to de novo metagenome assembly from short sequence reads. In: Proceedings of the 2nd ACM Conference on Bioinformatics, Computational Biology and Biomedicine, pp. 116–124.
Nasko, D.J., Chopyk, J., Sakowski, E.G., Ferrell, B.D., Polson, S.W., Wommack, K.E., 2018. Family A DNA polymerase phylogeny uncovers diversity and replication gene organization in the virioplankton. Front. Microbiol. 9, 3053.
Neuer, S., Cowles, T.J., 1994. Protist herbivory in the Oregon upwelling system. Marine Ecol. Prog. Ser. Oldendorf 113 (1), 147–162.
Nissimov, J.I., Jones, M., Napier, J.A., Munn, C.B., Kimmance, S.A., Allen, M.J., 2013. Functional inferences of environmental coccolithovirus biodiversity. Virol. Sin. 28 (5), 291–302.
Nissimov, J.I., Napier, J.A., Allen, M.J., Kimmance, S.A., 2016. Intragenus competition between coccolithoviruses: an insight on how a select few can come to dominate many. Environ. Microbiol. 18 (1), 133–145.
Nissimov, J.I., Pagarete, A., Ma, F., Cody, S., Dunigan, D.D., Kimmance, S.A., Allen, M.J., 2017. Coccolithoviruses: a review of cross-kingdom genomic thievery and metabolic thuggery. Viruses 9 (3), 52.

Nissimov, J.I., Vandzura, R., Johns, C.T., Natale, F., Haramaty, L., Bidle, K.D., 2018. Dynamics of transparent exopolymer particle production and aggregation during viral infection of the coccolithophore, *Emiliania huxleyi*. Environ. Microbiol. 20 (8), 2880–2897.

Nissimov, J.I., Talmy, D., Haramaty, L., Fredricks, H.F., Zelzion, E., Knowles, B., Eren, A.M., Vandzura, R., Laber, C.P., Schieler, B.M., Johns, C.T., More, K.D., Coolen, M.J.L., Follows, M.J., Bhattacharya, D., Van Mooy, B.A.S., Bidle, K.D., 2019a. Biochemical diversity of glycosphingolipid biosynthesis as a driver of Coccolithovirus competitive ecology. Environ. Microbiol. 21 (6), 2182–2197.

Nissimov, J.I., Talmy, D., Haramaty, L., Fredricks, H.F., Zelzion, E., Knowles, B., Eren, A.M., Vandzura, R., Laber, C.P., Schieler, B.M., Johns, C.T., 2019b. Biochemical diversity of glycosphingolipid biosynthesis as a driver of Coccolithovirus competitive ecology. Environ. Microbiol. 21 (6), 2182–2197.

Nurk, S., Meleshko, D., Korobeynikov, A., Pevzner, P.A., 2017. metaSPAdes: a new versatile metagenomic assembler. Genome Res. 27 (5), 824–834.

O'Neil, J.M., Davis, T.W., Burford, M.A., Gobler, C.J., 2012. The rise of harmful cyanobacteria blooms: the potential roles of eutrophication and climate change. Harmful Algae 14, 313–334.

Ohki, K., 1999. A possible role of temperate phage in the regulation of Trichodesmium. Bull. Inst. Océanogr. Monaco 19, 235–256.

Ortmann, A.C., Lawrence, J.E., Suttle, C.A., 2002. Lysogeny and lytic viral production during a bloom of the cyanobacterium *Synechococcus* spp. Microb. Ecol., 225–231.

Paasche, E., 2001. A review of the coccolithophorid *Emiliania huxleyi* (Prymnesiophyceae), with particular reference to growth, coccolith formation, and calcification-photosynthesis interactions. Phycologia 40 (6), 503–529.

Paerl, H.W., 2014. Mitigating harmful cyanobacterial blooms in a human-and climatically-impacted world. Life 4 (4), 988–1012.

Paerl, H.W., Huisman, J., 2009. Climate change: a catalyst for global expansion of harmful cyanobacterial blooms. Environ. Microbiol. Rep. 1 (1), 27–37.

Pahlow, M., Chien, C.T., Arteaga, L.A., Oschlies, A., 2020. Optimality-based non-Redfield plankton–ecosystem model (OPEM v1. 1) in UVic-ESCM 2.9–Part 1: implementation and model behaviour. Geosci. Model Dev. 13 (10), 4663–4690.

Parsons, R.J., Breitbart, M., Lomas, M.W., Carlson, C.A., 2012. Ocean time-series reveals recurring seasonal patterns of virioplankton dynamics in the northwestern Sargasso Sea. ISME J. 6 (2), 273–284.

Partensky, F., Blanchot, J., Lantoine, F., Neveux, J., Marie, D., 1996. Vertical structure of picophytoplankton at different trophic sites of the tropical northeastern Atlantic Ocean. Deep-Sea Res. I Oceanogr. Res. Pap. 43 (8), 1191–1213.

Pasulka, A.L., Thamatrakoln, K., Kopf, S.H., Guan, Y., Poulos, B., Moradian, A., Sweredoski, M.J., Hess, S., Sullivan, M.B., Bidle, K.D., Orphan, V.J., 2018. Interrogating marine virus-host interactions and elemental transfer with BONCAT and nanoSIMS-based methods. Environ. Microbiol. 20 (2), 671–692.

Pelusi, A., De Luca, P., Manfellotto, F., Thamatrakoln, K., Bidle, K.D., Montresor, M., 2021. Virus-induced spore formation as a defense mechanism in marine diatoms. New Phytol. 229 (4), 2251–2259.

Peng, Y., Leung, H.C., Yiu, S.M., Chin, F.Y., 2012. IDBA-UD: a de novo assembler for single-cell and metagenomic sequencing data with highly uneven depth. Bioinformatics 28 (11), 1420–1428.

Penta, W.B., Fox, J., Halsey, K.H., 2021. Rapid photoacclimation during episodic deep mixing augments the biological carbon pump. Limnol. Oceanogr. 66 (5), 1850–1866.

Petrou, K., Baker, K.G., Nielsen, D.A., Hancock, A.M., Schulz, K.G., Davidson, A.T., 2019. Acidification diminishes diatom silica production in the Southern Ocean. Nat. Clim. Chang. 9 (10), 781–786.

Piedade, G.J., Wesdorp, E.M., Montenegro-Borbolla, E., Maat, D.S., Brussaard, C.P., 2018. Influence of irradiance and temperature on the virus MpoV-45T infecting the Arctic picophytoplankter *Micromonas polaris*. Viruses 10 (12), 676.

Pinhassi, J., Zweifel, U.L., Hagstroëm, A., 1997. Dominant marine bacterioplankton species found among colony-forming bacteria. Appl. Environ. Microbiol. 63 (9), 3359–3366.

Pohlner, M., Degenhardt, J., von Hoyningen-Huene, A.J., Wemheuer, B., Erlmann, N., Schnetger, B., Badewien, T.H., Engelen, B., 2017. The biogeographical distribution of benthic Roseobacter group members along a Pacific transect is structured by nutrient availability within the sediments and primary production in different oceanic provinces. Front. Microbiol. 8, 2550.

Pomeroy, L.R., 1974. The ocean's food web, a changing paradigm. Bioscience 24 (9), 499–504.

Poorvin, L., Rinta-Kanto, J.M., Hutchins, D.A., Wilhelm, S.W., 2004. Viral release of iron and its bioavailability to marine plankton. Limnol. Oceanogr. 49 (5), 1734–1741.

Poullain, V., Gandon, S., Brockhurst, M.A., Buckling, A., Hochberg, M.E., 2008. The evolution of specificity in evolving and coevolving antagonistic interactions between a bacteria and its phage. Evol. Int. J. Organic Evol. 62 (1), 1–11.

Pourtois, J., Tarnita, C.E., Bonachela, J.A., 2020. Impact of lytic phages on phosphorus-vs. nitrogen-limited marine microbes. Front. Microbiol. 11, 221.

Proctor, L.M., Fuhrman, J.A., 1990. Viral mortality of marine bacteria and cyanobacteria. Nature 343 (6253), 60–62.

Puxty, R.J., Millard, A.D., Evans, D.J., Scanlan, D.J., 2015. Shedding new light on viral photosynthesis. Photosynth. Res. 126 (1), 71–97.

Qiu, Y., Gu, L., Tian, S., Sidhu, J., Gibbons, J., Van Den Top, T., Gonzalez-Hernandez, J.L., Zhou, R., 2019. Developmentally regulated genome editing in terminally differentiated N2-fixing heterocysts of Anabaena cylindrica ATCC. bioRxiv 29414, 629832.

Raina, J.B., Tapiolas, D., Willis, B.L., Bourne, D.G., 2009. Coral-associated bacteria and their role in the biogeochemical cycling of sulfur. Appl. Environ. Microbiol. 75 (11), 3492–3501.

Rappé, M.S., Connon, S.A., Vergin, K.L., Giovannoni, S.J., 2002. Cultivation of the ubiquitous SAR11 marine bacterioplankton clade. Nature 418 (6898), 630–633.

Richards, K.J., 2017. Viral infections of oceanic plankton blooms. J. Theor. Biol. 412, 27–35.

Richardson, A.J., Schoeman, D.S., 2004. Climate impact on plankton ecosystems in the Northeast Atlantic. Science 305 (5690), 1609–1612.

Ridgwell, A., Schmidt, D.N., Turley, C., Brownlee, C., Maldonado, M.T., Tortell, P., Young, J.R., 2009. From laboratory manipulations to Earth system models: scaling calcification impacts of ocean acidification. Biogeosciences 6 (11), 2611–2623.

Riebesell, U., Gattuso, J.P., 2015. Lessons learned from ocean acidification research. Nat. Clim. Chang. 5 (1), 12–14.

Riebesell, U., Bach, L.T., Bellerby, R.G., Monsalve, J., Boxhammer, T., Czerny, J., Larsen, A., Ludwig, A., Schulz, K.G., 2017. Competitive fitness of a predominant pelagic calcifier impaired by ocean acidification. Nat. Geosci. 10 (1), 19–23.

Rihtman, B., Puxty, R.J., Hapeshi, A., Lee, Y.-J., Zhan, Y., Michniewski, S., Waterfield, N.R., Chen, F., Weigele, P., Millard, A.D., Scanlan, D.J., Chen, Y., 2021. A new family of globally distributed lytic roseophages with unusual deoxythymidine to deoxyuridine substitution. Curr. Biol. 31 (14), 3199–3206.

Rimon, A., Oppenheim, A.B., 1975. Heat induction of the blue-green alga Plectonema boryanum lysogenic for the cyanophage SPIcts1. Virology 64 (2), 454–463.

Rohwer, F., Segall, A., Steward, G., Seguritan, V., Breitbart, M., Wolven, F., Farooq Azam, F., 2000. The complete genomic sequence of the marine phage Roseophage SIO1 shares homology with nonmarine phages. Limnol. Oceanogr. 45 (2), 408–418.

Rong, C., Zhou, K., Li, S., Xiao, K., Xu, Y., Zhang, R., Yang, Y., Zhang, Y., 2022. Isolation and characterization of a novel cyanophage encoding multiple auxiliary metabolic genes. Viruses 14 (5), 887.

Roossinck, M.J., Bazán, E.R., 2017. Symbiosis: viruses as intimate partners. Annu. Rev. Virol. 4 (1), 123–139.

Rose, S.L., Fulton, J.M., Brown, C.M., Natale, F., Van Mooy, B.A., Bidle, K.D., 2014. Isolation and characterization of lipid rafts in *Emiliania huxleyi*: a role for membrane microdomains in host–virus interactions. Environ. Microbiol. 16 (4), 1150–1166.

Rosenwasser, S., Mausz, M.A., Schatz, D., Sheyn, U., Malitsky, S., Aharoni, A., Weinstock, E., Tzfadia, O., Ben-Dor, S., Feldmesser, E., Pohnert, G., Vardi, A., 2014. Rewiring host lipid metabolism by large viruses determines the fate of *Emiliania huxleyi*, a bloom-forming alga in the ocean. Plant Cell 26 (6), 2689–2707.

Rosenwasser, S., Ziv, C., Van Creveld, S.G., Vardi, A., 2016. Virocell metabolism: metabolic innovations during host–virus interactions in the ocean. Trends Microbiol. 24 (10), 821–832.

Rost, B., Riebesell, U., 2004. Coccolithophores and the biological pump: responses to environmental changes. In: Coccolithophores. Springer, Berlin, Heidelberg, pp. 99–125.

Roux, S., Adriaenssens, E.M., Dutilh, B.E., Koonin, E.V., Kropinski, A.M., Krupovic, M., Kuhn, J.H., Lavigne, R., Brister, J.R., Varsani, A., Amid, C., Aziz, R.K., Bordenstein, S.R., Bork, P., Breitbart, M., Cochrane, G., Daly, R.A., Desnues, C., Duhaime, M.B., Emerson, J.B., Enault, F., Fuhrman, J.A., Pascal Hingamp, P., Hugenholtz, B., Hurwitz, B.L., Ivanova, N.N., Labonté, J.M., Lee, K.-B., Malmstrom, R.R., Martínez-García, M., Mizrachi, I.K., Ogata, H., Páez-Espino, D., Petit, M.-A., Putonti, C., Rattei, T., Reyes, A., Rodríguez-Valera, F., Rosario, K., Schriml, L.M., Schulz, F., Steward, G.F., Sullivan, M.B., Shinichi Sunagawa, C., Suttle, B.T., Tringe, S.G., Thurber, R.V., Webster, N.S., Whiteson, K.L., Wilhelm, S.W., Eric Wommack, K., Woyke, T., Wrighton, K.C., Yilmaz, P., Yoshida, T., Young, M.J., Yutin, N., Allen, L.Z., Kyrpides, N.C., Eloe-Fadrosh, E.A., 2019. Minimum Information about an Uncultivated Virus Genome (MIUViG). Nat. Biotechnol. 37, 29–37.

Ruardij, P., Veldhuis, M.J., Brussaard, C.P., 2005. Modeling the bloom dynamics of the polymorphic phytoplankter *Phaeocystis globosa*: impact of grazers and viruses. Harmful Algae 4 (5), 941–963.

Rynearson, T.A., Richardson, K., Lampitt, R.S., Sieracki, M.E., Poulton, A.J., Lyngsgaard, M.M., Perry, M.J., 2013. Major contribution of diatom resting spores to vertical flux in the sub-polar North Atlantic. Deep-Sea Res. I Oceanogr. Res. Pap. 82, 60–71.

Safferman, R.S., Morris, M.E., 1963. Algal virus: isolation. Science 140 (3567), 679–680.

Safferman, R.S., Morris, M.E., 1964. Growth characteristics of the blue-green algal virus LPP-1. J. Bacteriol. 88 (3), 771–775.

Sakowski, E.G., Munsell, E.V., Hyatt, M., Kress, W., Williamson, S.J., Nasko, D.J., Polson, S.W., Wommack, K.E., 2014. Ribonucleotide reductases reveal novel viral diversity and predict biological and ecological features of unknown marine viruses. Proc. Natl. Acad. Sci. 111 (44), 15786–15791.

Sandaa, R.A., Clokie, M., Mann, N.H., 2008. Photosynthetic genes in viral populations with a large genomic size range from Norwegian coastal waters. FEMS Microbiol. Ecol. 63 (1), 2–11.

Schaum, C.E., Collins, S., 2014. Plasticity predicts evolution in a marine alga. Proc. R. Soc. B Biol. Sci. 281 (1793), 20141486.

Schieler, B.M., Soni, M.V., Brown, C.M., Coolen, M.J., Fredricks, H., Van Mooy, B.A., Hirsh, D.J., Bidle, K.D., 2019. Nitric oxide production and antioxidant function during viral infection of the coccolithophore *Emiliania huxleyi*. ISME J. 13 (4), 1019–1031.

Schmidt, T.M. (Ed.), 2019. Encyclopedia of Microbiology. Academic Press.

Schmidt, H.F., Sakowski, E.G., Williamson, S.J., Polson, S.W., Wommack, K., 2014. Shotgun metagenomics indicates novel family A DNA polymerases predominate within marine virioplankton. ISME J. 8 (1), 103–114.

Schoemann, V., Becquevort, S., Stefels, J., Rousseau, V., Lancelot, C., 2005. *Phaeocystis* blooms in the global ocean and their controlling mechanisms: a review. J. Sea Res. 53 (1-2), 43–66.

Schroeder, D.C., Oke, J., Malin, G., Wilson, W.H., 2002. Coccolithovirus (Phycodnaviridae): characterisation of a new large dsDNA algal virus that infects *Emiliana huxleyi*. Arch. Virol. 147 (9), 1685–1698.

Schwalbach, M.S., Tripp, H.J., Steindler, L., Smith, D.P., Giovannoni, S.J., 2010. The presence of the glycolysis operon in SAR11 genomes is positively correlated with ocean productivity. Environ. Microbiol. 12 (2), 490–500.

Seager, R., Cane, M., Henderson, N., Lee, D.E., Abernathey, R., Zhang, H., 2019. Strengthening tropical Pacific zonal sea surface temperature gradient consistent with rising greenhouse gases. Nat. Clim. Chang. 9 (7), 517–522.

Selje, N., Simon, M., Brinkhoff, T., 2004. A newly discovered Roseobacter cluster in temperate and polar oceans. Nature 427 (6973), 445–448.

Shan, J., Jia, Y., Clokie, M.R., Mann, N.H., 2008. Infection by the 'photosynthetic'phage S-PM2 induces increased synthesis of phycoerythrin in *Synechococcus* sp. WH7803. FEMS Microbiol. Lett. 283 (2), 154–161.

Shang, S.Y., Ma, H., Zhao, Y.J., Cheng, K., 2016. Effect of nutrient status on the kinetics of cyanophage PP infection in Phormidium. *Ying Yong Sheng tai xue bao= The*. J. Appl. Ecol. 27 (4), 1271–1276.

Sheik, A.R., Brussaard, C.P.D., Lavik, G., Foster, R.A., Musat, N., Adam, B., Kuypers, M.M.M., 2013. Viral infection of *Phaeocystis globosa* impedes release of chitinous star-like structures: quantification using single cell approaches. Environ. Microbiol. 15 (5), 1441–1451.

Sheyn, U., Rosenwasser, S., Ben-Dor, S., Porat, Z., Vardi, A., 2016. Modulation of host ROS metabolism is essential for viral infection of a bloom-forming coccolithophore in the ocean. ISME J. 10 (7), 1742–1754.

Sheyn, U., Rosenwasser, S., Lehahn, Y., Barak-Gavish, N., Rotkopf, R., Bidle, K.D., Koren, I., Schatz, D., Vardi, A., 2018. Expression profiling of host and virus during a coccolithophore bloom provides insights into the role of viral infection in promoting carbon export. ISME J. 12 (3), 704–713.

Shirai, Y., Tomaru, Y., Takao, Y., Suzuki, H., Nagumo, T., Nagasaki, K., 2008. Isolation and characterization of a single-stranded RNA virus infecting the marine planktonic diatom *Chaetoceros tenuissimus* Meunier. Appl. Environ. Microbiol. 74 (13), 4022–4027.

Shitrit, D., Hackl, T., Laurenceau, R., Raho, N., Carlson, M.C., Sabehi, G., Schwartz, D.A., Chisholm, S.W., Lindell, D., 2022. Genetic engineering of marine cyanophages reveals integration but not lysogeny in T7-like cyanophages. ISME J. 16 (2), 488–499.

Silvano, A., Rintoul, S.R., Peña-Molino, B., Hobbs, W.R., van Wijk, E., Aoki, S., Tamura, T., Williams, G.D., 2018. Freshening by glacial meltwater enhances melting of ice shelves and reduces formation of Antarctic Bottom Water. Sci. Adv. 4 (4), eaap9467.

Singh, B.K., Chattopadhyay, J., Sinha, S., 2004. The role of virus infection in a simple phytoplankton zooplankton system. J. Theor. Biol. 231 (2), 153–166.

Slagter, H.A., Gerringa, L.J., Brussaard, C.P., 2016. Phytoplankton virus production negatively affected by iron limitation. Front. Mar. Sci. 3, 156.

Smayda, T.J., 2010. Adaptations and selection of harmful and other dinoflagellate species in upwelling systems. 2. Motility and migratory behaviour. Prog. Oceanogr. 85 (1-2), 71–91.

Smith, M.J., Tittensor, D.P., Lyutsarev, V., Murphy, E., 2015. Inferred support for disturbance-recovery hypothesis of North Atlantic phytoplankton blooms. J. Geophys. Res. Oceans 120 (10), 7067–7090.

Sonnenschein, E.C., Nielsen, K.F., D'Alvise, P., Porsby, C.H., Melchiorsen, J., Heilmann, J., Kalatzis, P.G., López-Pérez, M., Bunk, B., Spröer, C., Middelboe, M., Gram, L., 2017. Global occurrence and heterogeneity of the Roseobacter-clade species Ruegeria mobilis. ISME J. 11 (2), 569–583.

Steenhauer, L.M., Wierenga, J., Carreira, C., Limpens, R.W., Koster, A.J., Pollard, P.C., Brussaard, C.P., 2016. Isolation of cyanophage CrV infecting Cylindrospermopsis raciborskii and the influence of temperature and irradiance on CrV proliferation. Aquat. Microb. Ecol. 78 (1), 11–23.

Stellema, A., Sen Gupta, A., Taschetto, A.S., 2019. Projected slow down of South Indian Ocean circulation. Sci. Rep. 9 (1), 1–15.

Stock, C.A., Dunne, J.P., Fan, S., Ginoux, P., John, J., Krasting, J.P., Laufkötter, C., Paulot, F., Zadeh, N., 2020. Ocean biogeochemistry in GFDL's Earth System Model 4.1 and its response to increasing atmospheric CO2. J. Adv. Model. Earth Syst. 12 (10), e2019MS002043.

Stocker, T. (Ed.), 2014. Climate Change 2013: The Physical Science Basis: Working Group I Contribution to the Fifth Assessment Report of the Intergovernmental Panel on Climate Change. Cambridge University Press.

Stoddard, L.I., Martiny, J.B., Marston, M.F., 2007. Selection and characterization of cyanophage resistance in marine *Synechococcus* strains. Appl. Environ. Microbiol. 73 (17), 5516–5522.

Striebel, M., Schabhüttl, S., Hodapp, D., Hingsamer, P., Hillebrand, H., 2016. Phytoplankton responses to temperature increases are constrained by abiotic conditions and community composition. Oecologia 182 (3), 815–827.

Šulčius, S., Mazur-Marzec, H., Vitonytė, I., Kvederavičiūtė, K., Kuznecova, J., Šimoliūnas, E., Holmfeldt, K., 2018. Insights into cyanophage-mediated dynamics of nodularin and other non-ribosomal peptides in Nodularia spumigena. Harmful Algae 78, 69–74.

Sullivan, M.B., Waterbury, J.B., Chisholm, S.W., 2003. Cyanophages infecting the oceanic cyanobacterium *Prochlorococcus*. Nature 424 (6952), 1047–1051.

Sullivan, M.B., Lindell, D., Lee, J.A., Thompson, L.R., Bielawski, J.P., Chisholm, S.W., 2006. Prevalence and evolution of core photosystem II genes in marine cyanobacterial viruses and their hosts. PLoS Biol. 4 (8), e234.

Sullivan, M.B., Coleman, M.L., Weigele, P., Rohwer, F., Chisholm, S.W., et al., 2005. Three *Prochlorococcus* cyanophage genomes: Signature features and ecological interpretations. PLoS Biol. 3 (5), 144.

Sullivan, M.B., Huang, K.H., Ignacio-Espinoza, J.C., Berlin, A.M., Kelly, L., Weigele, P.R., DeFrancesco, A.S., Kern, S.E., Thompson, L.R., Young, S., Yandava, C., 2010. Genomic analysis of oceanic cyanobacterial myoviruses compared with T4-like myoviruses from diverse hosts and environments. Environ. Microbiol. 12 (11), 3035–3056.

Sun, J., Steindler, L., Thrash, J.C., Halsey, K.H., Smith, D.P., Carter, A.E., Landry, Z.C., Giovannoni, S.J., 2011. One carbon metabolism in SAR11 pelagic marine bacteria. PLoS One 6 (8), e23973.

Sun, J., Todd, J.D., Thrash, J.C., Qian, Y., Qian, M.C., Temperton, B., Guo, J., Fowler, E.K., Aldrich, J.T., Nicora, C.D., Lipton, M.S., Smith, R.D., De Leenheer, P., Payne, S.H., Johnston, A.W.B., Davie-Martin, C.L., Halsey, K.H., Giovannoni, S.J., 2016. The abundant marine bacterium Pelagibacter simultaneously catabolizes dimethylsulfoniopropionate to the gases dimethyl sulfide and methanethiol. Nat. Microbiol. 1 (8), 1–5.

Suttle, C.A., 2005. Viruses in the sea. Nature 437 (7057), 356–361.

Takano, Y., Tomaru, Y., Nagasaki, K., et al., 2018. Visualization of a dinoflagellate-infecting virus HcDNAV and its infection process. Viruses 10 (10), 554.

Talmy, D., Beckett, S.J., Taniguchi, D.A., Brussaard, C.P., Weitz, J.S., Follows, M.J., 2019. An empirical model of carbon flow through marine viruses and microzooplankton grazers. Environ. Microbiol. 21 (6), 2171–2181.

Tang, K., Zong, R., Zhang, F., Xiao, N., Jiao, N., 2010. Characterization of the photosynthetic apparatus and proteome of Roseobacter denitrificans. Curr. Microbiol. 60 (2), 124–133.

Tang, K., Yang, Y., Lin, D., Li, S., Zhou, W., Han, Y., Liu, K., Jiao, N., 2016. Genomic, physiologic, and proteomic insights into metabolic versatility in Roseobacter clade bacteria isolated from deep-sea water. Sci. Rep. 6 (1), 1–12.

Tarutani, K., Nagasaki, K., Itakura, S., Yamaguchi, M., 2001. Isolation of a virus infecting the novel shellfish-killing dinoflagellate *Heterocapsa circularisquama*. Aquat. Microb. Ecol. 23 (2), 103–111.

Taucher, J., Bach, L.T., Prowe, A.E., Boxhammer, T., Kvale, K., Riebesell, U., 2022. Enhanced silica export in a future ocean triggers global diatom decline. Nature 605 (7911), 696–700.

Thamatrakoln, K., Talmy, D., Haramaty, L., Maniscalco, C., Latham, J.R., Knowles, B., Natale, F., Coolen, M.J., Follows, M.J., Bidle, K.D., 2019. Light regulation of coccolithophore host–virus interactions. New Phytol. 221 (3), 1289–1302.

Thomas, M.K., Kremer, C.T., Litchman, E., 2016. Environment and evolutionary history determine the global biogeography of phytoplankton temperature traits. Glob. Ecol. Biogeogr. 25 (1), 75–86.

Thompson, L.R., Zeng, Q., Kelly, L., Huang, K.H., Singer, A.U., Stubbe, J., Chisholm, S.W., 2011. Phage auxiliary metabolic genes and the redirection of cyanobacterial host carbon metabolism. Proc. Natl. Acad. Sci. 108 (39), E757–E764.

Tomabechi, K., 2010. Energy resources in the future. Energies 3 (4), 686–695.

Tomaru, Y., Katanozaka, N., Nishida, K., Shirai, Y., Tarutani, K., Yamaguchi, M., Nagasaki, K., 2004. Isolation and characterization of two distinct types of HcRNAV, a single-stranded RNA virus infecting the bivalve-killing microalga *Heterocapsa circularisquama*. Aquat. Microb. Ecol. 34 (3), 207–218.

Tomaru, Y., Kimura, K., Nagasaki, K., 2015. Marine Protist Viruses. Marine Protists.

Tomaru, Y., Kimura, K., Yamaguchi, H., 2014. Temperature alters algicidal activity of DNA and RNA viruses infecting *Chaetoceros tenuissimus*. Aquat. Microb. Ecol. 73 (2), 171–183.

Tsai, A.Y., Gong, G.C., Chao, C.F., 2016. Contribution of viral lysis and nanoflagellate grazing to bacterial mortality at surface waters and deeper depths in the coastal ecosystem of subtropical Western Pacific. Estuar. Coasts 39 (5), 1357–1366.

Tyrrell, T., Merico, A., 2004. *Emiliania huxleyi*: bloom observations and the conditions that induce them. In: Coccolithophores. Springer, Berlin, Heidelberg, pp. 75–97.

Våge, S., Storesund, J.E., Thingstad, T.F., 2013. SAR11 viruses and defensive host strains. Nature 499 (7459), E3–E4.

Vallina, S.M., Cermeno, P., Dutkiewicz, S., Loreau, M., Montoya, J.M., 2017. Phytoplankton functional diversity increases ecosystem productivity and stability. Ecol. Model. 361, 184–196.

Van Etten, J.L., Burbank, D.E., Xia, Y., Meints, R.H., 1983. Growth cycle of a virus, PBCV-1, that infects Chlorella-like algae. Virology 126 (1), 117–125.

Van Etten, J.L., Graves, M.V., Müller, D.G., Boland, W., Delaroque, N., 2002. Phycodnaviridae–large DNA algal viruses. Arch. Virol. 147 (8), 1479–1516.

van Kampen, N.G., 2007. Stochastic Processes in Physics and Chemistry, third ed. Elsevier.

Vardi, A., Haramaty, L., Van Mooy, B.A., Fredricks, H.F., Kimmance, S.A., Larsen, A., Bidle, K.D., 2012. Host–virus dynamics and subcellular controls of cell fate in a natural coccolithophore population. Proc. Natl. Acad. Sci. 109 (47), 19327–19332.

Vardi, A., Van Mooy, B.A., Fredricks, H.F., Popendorf, K.J., Ossolinski, J.E., Haramaty, L., Bidle, K.D., 2009. Viral glycosphingolipids induce lytic infection and cell death in marine phytoplankton. Science 326 (5954), 861–865.

Vargas, C.A., Martínez, R.A., Cuevas, L.A., Pavez, M.A., Cartes, C., González, H.E., Escribano, R., Daneri, G., 2007. The relative importance of microbial and classical food webs in a highly productive coastal upwelling area. Limnol. Oceanogr. 52, 1495–1510.

Vincent, F., Sheyn, U., Porat, Z., Vardi, A., et al., 2021. Visualizing active viral infection reveals diverse cell fates in synchronized algal bloom demise. Proc. Natl. Acad. Sci. 118 (11), e2021586118.

von Meijenfeldt, F.B., Hogeweg, P., Dutilh, B.E., 2022. On specialists and generalists: niche range strategies across the tree of life. bioRxiv.

Wagner-Döbler, I., Biebl, H., 2006. Environmental biology of the marine Roseobacter lineage. Ann. Rev. Microbiol. 60, 255–280.

Waldbauer, J.R., Coleman, M.L., Rizzo, A.I., Campbell, K.L., Lotus, J., Zhang, L., 2019. Nitrogen sourcing during viral infection of marine cyanobacteria. Proc. Natl. Acad. Sci. 116 (31), 15590–15595.

Warwick-Dugdale, J., Solonenko, N., Moore, K., Chittick, L., Gregory, A.C., Allen, M.J., Sullivan, M.B., Temperton, B., 2019. Long-read viral metagenomics captures abundant and microdiverse viral populations and their niche-defining genomic islands. PeerJ 7, e6800.

Webb, V., Leduc, E., Spiegelman, G.B., 1982. Burst size of bacteriophage SP82 as a function of growth rate of its host Bacillus subtilis. Can. J. Microbiol. 28 (11), 1277–1280.

Weitz, J.S., Hartman, H., Levin, S.A., 2005. Coevolutionary arms races between bacteria and bacteriophage. Proc. Natl. Acad. Sci. 102 (27), 9535–9540.

Weitz, J.S., Stock, C.A., Wilhelm, S.W., Bourouiba, L., Coleman, M.L., Buchan, A., Follows, M.J., Fuhrman, J.A., Jover, L.F., Lennon, J.T., Middelboe, M., 2015. A multitrophic model to quantify the effects of marine viruses on microbial food webs and ecosystem processes. ISME J. 9 (6), 1352–1364.

Westbroek, P., Brown, C.W., van Bleijswijk, J., Brownlee, C., Brummer, G.J., Conte, M., Egge, J., Fernández, E., Jordan, R., Knappertsbusch, M., Stefels, J., Veldhuis, M., van der Wal, P., Young, J., 1993. A model system approach to biological climate forcing: the example of *Emiliania huxleyi*. Glob. Planet. Chang. 8 (1-2), 27–46.

Wilhelm, S.W., Suttle, C.A., 1999. Viruses and nutrient cycles in the sea: viruses play critical roles in the structure and function of aquatic food webs. Bioscience 49 (10), 781–788.

Wilson, W.H., Carr, N.G., Mann, N.H., 1996. The effect of phosphate status on the kinetics of cyanophage infection in the oceanic cyanobacterium *Synechococcus* sp. wh7803 1. J. Phycol. 32 (4), 506–516.

Wilson, W.H., Fuller, N.J., Joint, I.R., Mann, N.H., 2000. Analysis of cyanophage diversity in the marine environment using denaturing gradient gel electrophoresis. In: Microbial Biosystems: New Frontiers. Proceedings of the 8th International Symposium on Microbial Ecology. Atlantic Canada Society for Microbial Ecology, Halifax, Nova Scotia, Canada, pp. 565–570.

Wilson, W.H., Schroeder, D.C., Allen, M.J., Holden, M.T.G., Parkhill, J., Barrell, B.G., Churcher, C., Hamlin, N., Mungall, K., Norbertczak, H., Quail, M.A., Price, C., Rabbinowitsch, E., Walker, D., Craigon, M., Roy, D., Ghazal, P., 2005. Complete genome sequence and lytic phase transcription profile of a Coccolithovirus. Science 309 (5737), 1090–1092.

Wilson, W.H., Turner, S., Mann, N.H., 1998. Population dynamics of phytoplankton and viruses in a phosphate-limited mesocosm and their effect on DMSP and DMS production. Estuar. Coast. Shelf Sci. 46 (2), 49–59.

Wittmers, F., Needham, D.M., Hehenberger, E., Giovannoni, S.J., Worden, A.Z., 2022. Genomes from uncultivated pelagiphages reveal multiple phylogenetic clades exhibiting extensive auxiliary metabolic genes and cross-family multigene transfers. Msystems. e01522-21.

Wommack, K.E., Bhavsar, J., Ravel, J., 2008. Metagenomics: read length matters. Appl. Environ. Microbiol. 74 (5), 1453–1463.
Wommack, K.E., Sime-Ngando, T., Winget, D.M., Jamindar, S., Helton, R.R., 2010. Filtration-based methods for the collection of viral concentrates from large water samples. Manual of Aquatic Viral Ecology (MAVE). Advancing the Science for Limnology and Oceanography (ASLO), pp. 110–117.
Worden, A.Z., Nolan, J.K., Palenik, B., 2004. Assessing the dynamics and ecology of marine picophytoplankton: the importance of the eukaryotic component. Limnol. Oceanogr. 49 (1), 168–179.
Wu, W., Zhu, Q., Liu, X., An, C., Wang, J., 2009. Isolation of a freshwater cyanophage (F1) capable of infecting Anabaena flos-aquae and its potentials in the control of water bloom. Int. J. Environ. Pollut. 38 (1-2), 212–221.
Xiao, X., Guo, W., Li, X., Wang, C., Chen, X., Lin, X., Weinbauer, M.G., Zeng, Q., Jiao, N., Zhang, R., 2021. Viral lysis alters the optical properties and biological availability of dissolved organic matter derived from *Prochlorococcus* picocyanobacteria. Appl. Environ. Microbiol. 87 (3), e02271-20.
Xie, L., Zhang, R., Luo, Y.W., 2022. Assessment of explicit representation of dynamic viral processes in regional marine ecological models. Viruses 14 (7), 1448.
Yadav, S., Ahn, Y.H., 2021. Growth characteristics of lytic cyanophages newly isolated from the Nakdong River, Korea. Virus Res. 306, 198600.
Yamada, Y., Tomaru, Y., Fukuda, H., Nagata, T., 2018. Aggregate formation during the viral lysis of a marine diatom. Front. Mar. Sci. 5, 167.
Yamamoto-Kawai, M., McLaughlin, F.A., Carmack, E.C., Nishino, S., Shimada, K., 2009. Aragonite undersaturation in the Arctic Ocean: effects of ocean acidification and sea ice melt. Science 326 (5956), 1098–1100.
Yang, Y., Cai, L., Ma, R., Xu, Y., Tong, Y., Huang, Y., Jiao, N., Zhang, R., 2017. A novel roseosiphophage isolated from the oligotrophic South China Sea. Viruses 9 (5), 109.
Yau, S., Lauro, F.M., DeMaere, M.Z., Brown, M.V., Thomas, T., Raftery, M.J., Andrews-Pfannkoch, C., Lewis, M., Hoffman, J.M., Gibson, J.A., Cavicchioli, R., 2011. Virophage control of Antarctic algal host–virus dynamics. Proc. Natl. Acad. Sci. 108 (15), 6163–6168.
You, L., Suthers, P.F., Yin, J., 2002. Effects of *Escherichia coli* physiology on growth of phage T7 in vivo and in silico. J. Bacteriol. 184 (7), 1888–1894.
Zaucker, F., Stocker, T.F., Broecker, W.S., 1994. Atmospheric freshwater fluxes and their effect on the global thermohaline circulation. J. Geophys. Res. Oceans 99 (C6), 12443–12457.
Zehr, J.P., 2011. Nitrogen fixation by marine cyanobacteria. Trends Microbiol. 19 (4), 162–173.
Zehr, J.P., Capone, D.G., 2020. Changing perspectives in marine nitrogen fixation. Science 368 (6492), eaay9514.
Zhang, D., You, F., He, Y., Te, S.H., Gin, K.Y.H., 2020. Isolation and characterization of the first freshwater cyanophage infecting Pseudanabaena. J. Virol. 94 (17), e00682-20.
Zhang, Y., Jiao, N., 2009. Roseophage RDJLΦ1, infecting the aerobic anoxygenic phototrophic bacterium Roseobacter denitrificans OCh114. Appl. Environ. Microbiol. 75 (6), 1745–1749.
Zhang, Y., Sun, Y., Jiao, N., Stepanauskas, R., Luo, H., 2016. Ecological genomics of the uncultivated marine Roseobacter lineage CHAB-I-5. Appl. Environ. Microbiol. 82 (7), 2100–2111.
Zhang, Z., Chen, F., Chu, X., Zhang, H., Luo, H., Qin, F., Zhai, Z., Yang, M., Sun, J., Zhao, Y., 2019a. Diverse, abundant, and novel viruses infecting the marine Roseobacter RCA lineage. Msystems 4 (6), e00494-19.
Zhang, Z., Qin, F., Chen, F., Chu, X., Luo, H., Zhang, R., Du, S., Tian, Z., Zhao, Y., 2019b. Novel pelagiphages prevail in the ocean. bioRxiv.

Zhang, Z., Qin, F., Chen, F., Chu, X., Luo, H., Zhang, R., Du, S., Tian, Z., Zhao, Y., 2021. Culturing novel and abundant pelagiphages in the ocean. Environ. Microbiol. 23 (2), 1145–1161.

Zhan, Y., Chen, F., 2019a. Bacteriophages that infect marine roseobacters: genomics and ecology. Environ. Microbiol. 21 (6), 1885–1895.

Zhan, Y., Chen, F., 2019b. The smallest ssDNA phage infecting a marine bacterium. Environ. Microbiol. 21 (6), 1916–1928.

Zhan, Y., Huang, S., Voget, S., Simon, M., Chen, F., 2016. A novel roseobacter phage possesses features of podoviruses, siphoviruses, prophages and gene transfer agents. Sci. Rep. 6 (1), 1–8.

Zhao, J., Jing, H., Wang, Z., Wang, L., Jian, H., Zhang, R., Xiao, X., Chen, F., Jiao, N., Zhang, Y., 2022. Novel viral communities potentially assisting in carbon, nitrogen, and sulfur metabolism in the upper slope sediments of Mariana trench. Msystems 7 (1), e01358-21.

Zhao, Y., Qin, F., Zhang, R., Giovannoni, S.J., Zhang, Z., Sun, J., Du, S., Rensing, C., 2019a. Pelagiphages in the Podoviridae family integrate into host genomes. Environ. Microbiol. 21 (6), 1989–2001.

Zhao, Y., Temperton, B., Thrash, J.C., Schwalbach, M.S., Vergin, K.L., Landry, Z.C., Ellisman, M., Deerinck, T., Sullivan, M.B., Giovannoni, S.J., et al., 2013. Abundant SAR11 viruses in the ocean. Nature 494 (7437), 357–360.

Zhao, Y., Wang, K., Ackermann, H.W., Halden, R.U., Jiao, N., Chen, F., 2010. Searching for a "hidden" prophage in a marine bacterium. Appl. Environ. Microbiol. 76 (2), 589–595.

Zhao, Y., Wang, K., Jiao, N., Chen, F., 2009. Genome sequences of two novel phages infecting marine roseobacters. Environ. Microbiol. 11 (8), 2055–2064.

Zhao, Z., Gonsior, M., Schmitt-Kopplin, P., Zhan, Y., Zhang, R., Jiao, N., Chen, F., 2019b. Microbial transformation of virus-induced dissolved organic matter from picocyanobacteria: coupling of bacterial diversity and DOM chemodiversity. ISME J. 13 (10), 2551–2565.

Zheng, Q., Lin, W., Wang, Y., Li, Y., He, C., Shen, Y., Guo, W., Shi, Q., Jiao, N., 2021. Highly enriched N-containing organic molecules of *Synechococcus* lysates and their rapid transformation by heterotrophic bacteria. Limnol. Oceanogr. 66 (2), 335–348.

Zimmerman, A.E., Bachy, C., Ma, X., Roux, S., Jang, H.B., Sullivan, M.B., Waldbauer, J.R., Worden, A.Z., 2019. Closely related viruses of the marine picoeukaryotic alga *Ostreococcus lucimarinus* exhibit different ecological strategies. Environ. Microbiol. 21 (6), 2148–2170.

Zimmerman, A.E., Howard-Varona, C., Needham, D.M., John, S.G., Worden, A.Z., Sullivan, M.B., Waldbauer, J.R., Coleman, M.L., 2020. Metabolic and biogeochemical consequences of viral infection in aquatic ecosystems. Nat. Rev. Microbiol. 18 (1), 21–34.

Zingone, A., Sarno, D., Forlani, G., 1999. Seasonal dynamics in the abundance of *Micromonas pusilla* (Prasinophyceae) and its viruses in the Gulf of Naples (Mediterranean Sea). J. Plankton Res. 21 (11), 2143–2159.

Ziv, C., Malitsky, S., Othman, A., Ben-Dor, S., Wei, Y., Zheng, S., Aharoni, A., Hornemann, T., Vardi, A., 2016. Viral serine palmitoyltransferase induces metabolic switch in sphingolipid biosynthesis and is required for infection of a marine alga. Proc. Natl. Acad. Sci. 113 (13), E1907–E1916.

CHAPTER THREE

West Nile virus and climate change

Rachel L. Fay[a,b], Alexander C. Keyel[a,c], and Alexander T. Ciota[a,b,*]
[a]The Arbovirus Laboratory, Wadsworth Center, New York State Department of Health, Slingerlands, NY, United States
[b]Department of Biomedical Sciences, State University of New York at Albany School of Public Health, Rensselaer, NY, United States
[c]Department of Atmospheric and Environmental Sciences, State University of New York at Albany, Albany, NY, United States
[*]Corresponding author: e-mail address: alexander.ciota@health.ny.gov

Contents

1. Introduction 148
2. Temperature, viral fitness and vector competence for West Nile virus 153
 2.1 Population and species-specific relationships between temperature and vector competence 156
 2.2 Mosquito genetics, immunity and climate 156
 2.3 Mosquito microbiome and climate 157
 2.4 Additional considerations for climate and vector competence 159
3. Mosquito physiology and climate 160
 3.1 Additional considerations with climate and *Culex* fitness 162
 3.2 Trait-based models for West Nile virus and temperature 163
4. Epidemiological models of West Nile virus and climate 164
 4.1 Temperature and West Nile virus 164
 4.2 The influence of precipitation, humidity, and soil moisture on West Nile virus 167
 4.3 Vector and host distribution 169
5. The influence of temperature on West Nile virus diversity and evolution 172
6. Concluding remarks 174
References 176
Further reading 193

Abstract

West Nile virus (WNV) is a mosquito-borne flavivirus with a global distribution that is maintained in an enzootic cycle between *Culex* species mosquitoes and avian hosts. Human infection, which occurs as a result of spillover from this cycle, is generally subclinical or results in a self-limiting febrile illness. Central nervous system infection occurs in a minority of infections and can lead to long-term neurological complications and, rarely, death. WNV is the most prevalent arthropod-borne virus in the United States. Climate change can influence several aspects of WNV transmission including the vector,

Advances in Virus Research, Volume 114
ISSN 0065-3527
https://doi.org/10.1016/bs.aivir.2022.08.002

amplifying host, and virus. Climate change is broadly predicted to increase WNV distribution and risk across the globe, yet there will likely be significant regional variability and limitations to this effect. Increases in temperature can accelerate mosquito and pathogen development, drive increases in vector competence for WNV, and also alter mosquito life history traits including longevity, blood feeding behavior and fecundity. Precipitation, humidity and drought also impact WNV transmissibility. Alteration in avian distribution, diversity and phenology resulting from climate variation add additional complexity to these relationships. Here, we review WNV epidemiology, transmission, disease and genetics in the context of laboratory studies, field investigations, and infectious disease models under climate change. We summarize how mosquito genetics, microbial interactions, host dynamics, viral strain, population size, land use and climate account for distinct relationships that drive WNV activity and discuss how these dynamic and evolving interactions could shape WNV transmission and disease under climate change.

1. Introduction

Flaviviruses (family *Flaviviridae*) are enveloped viruses with positive-sense RNA genomes encoding a single polypeptide comprised of three structural (PrM, M, E) and seven non-structural (NS1, 2a, 2b, 3, 4a, 4b, and 5) proteins that are co- and post-translationally cleaved by virus and host-derived proteases (Brinton, 2002). The flavivirus genus includes the arthropod-borne viruses (arboviruses) that are most commonly associated with human disease. These include viruses primarily vectored by *Aedes* mosquitoes which rely on humans or non-human primates as amplifying hosts such as dengue virus (DENV), Zika virus (ZIKV), and Yellow Fever virus, enzootic tick-borne viruses such as tick-borne encephalitis virus and Powassan virus, and enzootic viruses that are primarily vectored by *Culex* mosquitoes and rely on birds as amplifying hosts such as Japanese encephalitis virus, Murray Valley encephalitis virus, St. Louis encephalitis virus and West Nile virus (WNV). These flaviviruses, particularly DENV, ZIKV and WNV, have experienced dramatic expansions over the last four decades, and WNV is now the most geographically widespread arboviruses in the world (Fig. 1).

WNV was first isolated in 1937 from a febrile woman in the West Nile district of Uganda (Smithburn et al., 1940). Infections were reported throughout the 1940s and 1950s in Egypt and Israel and associated with West Nile fever, a generally self-limiting febrile illness (Bernkopf et al., 1953; Goldblum et al., 1954; Melnick et al., 1951; Smithburn et al., 1940). The first documented outbreak of WNV occurred in a settlement

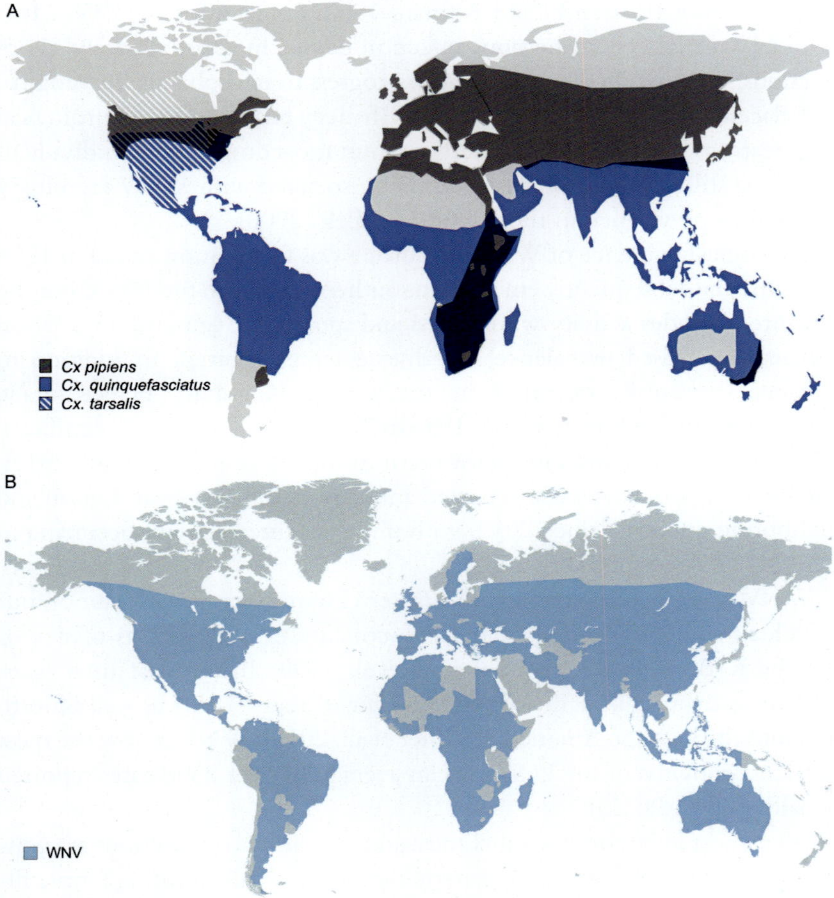

Fig. 1 The global distribution of West Nile virus and its primary vectors. West Nile virus (WNV) is vectored by *Culex* spp. mosquitoes. (A) The primary vectors include *Cx. pipiens* (gray), *Cx. quinquefasciatus* (blue), and, in the Americas, *Cx. tarsalis* (hashed). *Cx. pipiens* and *Cx. quinquefasicatus* readily hybridize in areas where they co-exist (dark blue). (B) WNV has been identified on all continents except Antarctica. Regions with evidence of local transmission (human disease, isolation from vector or host or serological evidence) are shown. WNV distribution generally coincides with the establishment of primary vectors.

near Haifa, Israel in 1951, with infection identified in 123 of 303 residents (Bernkopf et al., 1953). Symptoms of West Nile fever including myalgia, arthralgia, headache, rash and abdominal pain, were defined during this initial outbreak investigation. The potential for WNV to cause neurological complications was first noted in elderly Israeli patients in the late 1950s and early 1960s (Pruzanski and Altman, 1962; Spigland et al., 1958). It is now known that central nervous infection occurs in approximately 1% of infections, and that these infections can progress to encephalitis, meningitis, acute flaccid paralysis and, rarely, death (Chancey et al., 2015). Neurological complications largely occur in elderly or immuno-compromised individuals and, while still rare, neuroinvasive disease associated with WNV seemingly increased in prevalence in the 1990s (Hubálek, 2000).

Serological evidence of WNV in Europe was first demonstrated in 1958 in Albania and subsequently emerged in southern France in the 1960s. For the next three decades activity was limited and sporadic. Beginning in 1996 in Romania, increased prevalence and disease were reported. In addition to Romania, outbreaks occurred in the Czech Republic, France, Italy, Hungary, Spain and Portugal from 1996 to 2009. From 2010 to 12, significant WNV outbreaks occurred in southeastern Europe (Paz and Semenza, 2013). Hundreds of cases have been reported annually in the European Union and neighboring countries since 2012, with unprecedented activity occurring in 2018 (>2000 cases; ECDC, 2022).

WNV was first identified in the Western Hemisphere in a cluster of four individuals in New York City in 1999, concurrent with reports of sick or dying birds (Briese et al., 1999; Lanciotti et al., 1999). In a span of just 4 years, WNV spread throughout the contiguous United States (U.S.), as well as north and south through the Americas (Kramer et al., 2019). WNV is now the most prevalent arbovirus in the U.S, with an average of over 2500 cases reported annually (CDC, 2022a).

Since most infections are subclinical and symptomatic infections are generally associated with relatively generic symptoms of self-limiting febrile illness that remain undiagnosed, WNV cases are vastly underreported. It is estimated that there are approximately 150 infections for every reported case (Hayes et al., 2005). Among reported cases, which are largely neuroinvasive, mortality rates are ~5% (CDC, 2022b). There are currently no WNV vaccines or specific therapeutics licensed for use in humans, although approved equine vaccines are available (Saiz, 2020).

Epidemiological investigations in Egypt in the early 1950s established that WNV was maintained in an enzootic cycle between primarily *Culex* species mosquitoes and birds (Taylor et al., 1956; Taylor and Hurlbut, 1953) (Fig. 2). WNV is a generalist, with isolation and laboratory transmission reported in a wide range of mosquito species, with *Ae. aegypti* a notable exception (Ciota et al., 2017). WNV has additionally been isolated from hundreds of bird species, with passerine songbirds playing a primary role

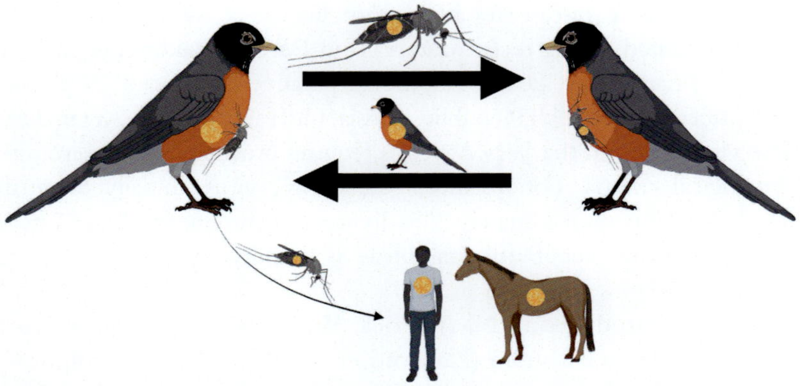

Fig. 2 The primary transmission cycle of West Nile virus (WNV). WNV is maintained in nature in an enzootic cycle involving primarily bird biting *Culex* spp. mosquitoes and avian hosts. Infection and disease in humans and other mammals occur as a result of spillover from this cycle.

in virus amplification (Kilpatrick and Wheeler, 2019). Humans, horses and other mammals are generally dead-end hosts that produce insufficient viremia for mosquito reinfection regardless of disease manifestations (Hayes et al., 2005). The primary vectors in nature are members of the bird biting *Culex pipiens L. complex*, most commonly *Cx. pipiens* and *Cx. quinquefasciatus*, which readily hybridize where ranges intersect (Farajollahi et al., 2011). *Cx. tarsalis*, which played a pivotal role in the western movement of WNV in the U.S., are also a primary vector in the western and southwestern U.S. (Goldberg et al., 2010; Fig. 1). As both species and population-specific differences in competence and overall vectorial capacity are well-documented (Ciota and Kramer, 2013), extrinsic factors that influence distributions are likely to have significant impacts on transmissibility and disease.

WNV can be separated into as many as seven genetically distinct lineages (Bakonyi et al., 2006; Ciota and Kramer, 2013) yet if a criterion of 20% genetic divergence is employed (Charrel et al., 2003) there are just two primary distinct lineages. Lineage 1 is the most widely distributed, with occurrence in Africa, Asia, Europe, Australia, and the Americas. A single point introduction of Lineage 1a was responsible for commencement of WNV activity in the Americas. Lineage 1b, also known as Kunjin virus, is found exclusively in Australia, and Lineage 1c is unique to India (Bondre et al., 2007). Lineage 2 originated in sub-Saharan Africa and has been responsible for a number of outbreaks in Europe in recent years (Hernández-Triana et al., 2014). Although previously thought to be associated exclusively with West Nile fever, neuroinvasive disease resulting from lineage 2 WNV infection has been demonstrated during recent outbreaks in Europe (Bakonyi et al., 2013; Hernández-Triana et al., 2014). Additional putative lineages of WNV have been identified in the Czech Republic, Russia, India and Malaysia, yet inclusion of these virus strains in the WNV species requires broader genetic, serological and ecological criteria. For instance, Rabensburg virus, initially designated lineage 3 WNV, is approximately 25% divergent from lineage 1 and 2 genotypes and is incapable of utilizing birds as amplifying hosts (Aliota and Kramer, 2012).

The strain introduced to N. America, WNV NY99, was most closely related to a 1998 Israeli strain (Lanciotti et al., 1999). WNV evolution in the U.S. has been dominated by purifying selection, with little broad geographic structure and a general lack of widespread fixed, consensus mutation. Despite this, notable genetic change has occurred. The introduced strain possessed a nonsynonymous mutation in the viral helicase that is associated with increased fitness and virulence in birds (Brault et al., 2007). In 2002, the WN02 genotype emerged and rapidly displaced NY99 (Davis et al., 2005). This genotype, which is characterized by a shared amino acid substitution in the WNV envelope, was subsequently shown to be more transmissible by *Culex* mosquitoes, particularly *Cx. tarsalis* (Moudy et al., 2007). In 2003, an additional genotype SW/WN03 was identified (McMullen et al., 2011). More recently, novel genotypes have been identified in New York State (Bialosuknia et al., 2022). WNV NY10 genotype strains, characterized by single, positively selected amino acid substitutions in the NS2a and NS4b, are associated with increased WNV prevalence and have been shown to be more transmissible by *Culex* mosquitoes and birds (Bialosuknia et al., 2022). These data demonstrate that adaptive evolution in the U.S. in ongoing.

Long-term alterations to global climate, resulting primarily from anthropogenic activities increasing greenhouse gas emissions, continue to accelerate globally. The intergovernmental panel on climate change (IPCC) has reported a 1.1 °C increase in global temperatures since the mid-1800s, with accelerating warming noted in each of the last four decades relative to the previous decade (Pachauri et al., 2014). In addition to temperature, climate change influences precipitation, sea level and acidification, and extreme weather events. While the impacts of climate change on public health are far-reaching, the biology and ecology of vector-borne pathogens are uniquely influenced by environmental variability. Climate change is therefore predicted to disproportionally influence vector-borne disease transmission (Pachauri et al., 2014; Watts et al., 2021).

The invasion of WNV in the Americas and other regions can be proximately explained by globalization, and the highly successful establishment and spread were facilitated by the presence of widespread populations of relatively competent vectors and naïve amplifying hosts, yet WNV transmission is inextricably bound to climate. Temperature, precipitation and drought have all been demonstrated to influence WNV prevalence and disease. Unraveling the role of climate in vector-borne pathogen transmission requires consideration of the complexity of vector-virus-host interactions. In this chapter we review what is known about how dynamic environments influence these interactions and how this knowledge can inform our understanding of the trajectory of WNV under climate change.

2. Temperature, viral fitness and vector competence for West Nile virus

Due to the need to cycle between biologically and environmentally disparate hosts, arboviruses inherently possess relatively high levels of phenotypic robustness (Lauring et al., 2013). This intrinsic capacity to retain fitness in distinct environments is likely to increase the capacity for WNV and other arboviruses to thrive under rapidly changing climates. Despite these relatively high levels of canalization, temperature variation does directly impact viral fitness and vector competence for arboviruses. Studies characterizing the impact of temperature on arboviruses and their interactions with their vectors can begin to shed some light on the changing transmission dynamics of WNV and other arboviruses in a warming globe.

Vector competence refers to the intrinsic capacity of a mosquito to transmit a pathogen following exposure (Fig. 3). In mosquitoes, this process begins with establishment of infection in midgut epithelial cells following blood meal ingestion. The efficacy of viral infection is initially governed by the binding of viral surface proteins with specific host cell receptors, and these interactions are known to be temperature sensitive (Gale, 2020). While WNV receptors have not been fully characterized and direct impacts of temperature on receptor binding has not been explicitly studied, experimental infections can provide insight into the relationship between infectivity and temperature. WNV is thought to infect at a minimum ambient temperature of $\sim 15\,^{\circ}C$ in mosquitoes (Shocket et al., 2020). Although negative impacts on mosquito survival preclude the capacity to study

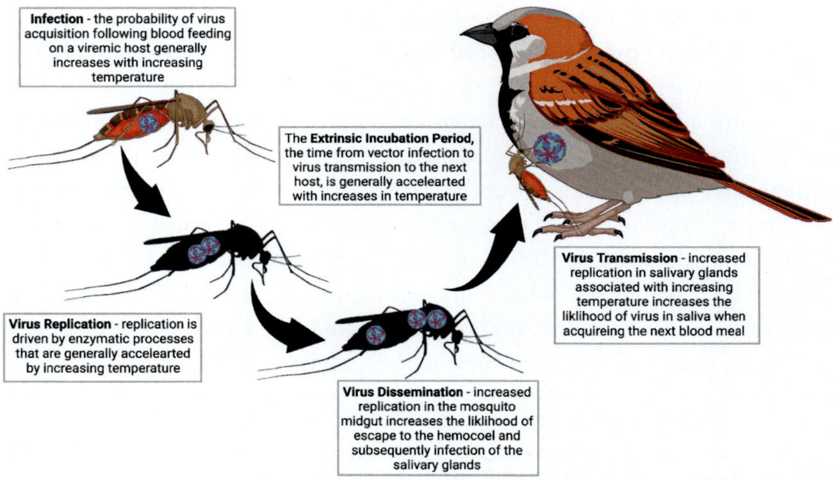

Fig. 3 The influence of temperature on vector competence for West Nile virus (WNV). Increased temperature has been shown to accelerate extrinsic incubation periods in *Culex* spp. mosquitoes and effect WNV infection, dissemination and transmission. Generic effects are shown but the relationship between temperature and vector competence is variable among unique species, mosquito populations and WNV strains.

impacts on infection at higher temperature (greater than ~34 °C), the fact that WNV has been shown to replicate in avian hosts at temperatures up to 44 °C (Kinney et al., 2006) demonstrates there is unlikely to be a significant constraint on infection imposed by rising ambient temperature that would limit viral fitness in mosquitoes. Indeed, experimental studies of WNV in mosquitoes have generally identified a wide range of suitable temperatures for infection and an increase of stable infection rates with increasing temperatures [reviewed in Ciota and Keyel (2019)].

Replication of WNV and other related viruses is controlled by a series of enzymatic reactions facilitated by components of the viral replication complex including the RNA-dependent RNA polymerase (RdRp), viral helicase, ATPase, protease, methyltransferase and guanyltransferase (Caldwell et al., 2022). While the non-structural genes NS3 and NS5 encode for these enzymes, activity is also facilitated by host proteins and influenced by interactions with additional non-structural viral proteins in the replication complex (Caldwell et al., 2022). The optimal conditions for enzymatic function are certainly variable among individual proteins and among WNV strains (Andrade et al., 2011; Dietrich et al., 2016; Ngo et al., 2019), yet increasing ambient temperatures generally accelerate enzymatic processes, consistent with the fact that increased replication rates are generally measured at higher temperatures for WNV and other viruses. The primary phenotypic consequences of increased replication rates for mosquito-borne viruses are both an increase in dissemination rates (the proportion of mosquitoes for which viral infections escape the gut) and a decrease in the extrinsic incubation period (EIP). The EIP refers to the length of time from virus infection to virus transmission by a vector. Arboviruses escape the midgut following replication in epithelial cells by traversing through gaps in the basal lamina into the hemocoel (Hardy et al., 1983). The likelihood of virus escape is directly correlated to viral load (Ciota et al., 2011b) and thresholds for escape are generally achieved more quickly with increased replication as a result of higher temperatures. This increased dissemination and shortened EIP significantly increases overall transmissibility, both because virus disseminated into the mosquito hemolymph will ultimately result in more infection in salivary glands, and because deceased EIP will increase the likelihood that a mosquito will be transmitting when seeking its next blood meal. Increased replication in salivary glands will additionally increase the likelihood of expelling virus into saliva, subsequently resulting in higher transmission rates (Fig. 3).

A number of experimental studies with WNV and its vectors have confirmed that viral load and overall vector competence are directly correlated to temperature, and that there is an inverse relationship between EIP and temperature (Fig. 3) (Ciota and Keyel, 2019). Although most studies have been carried out using constant temperature regimes, studies utilizing diurnal cycles that more accurately depict natural conditions by and large demonstrate that competence at a constant temperature is generally representative of cycling temperature around that mean (Cornel et al., 1993; Danforth et al., 2016). As temperature variability is predicted to be more disparate under climate change, both for daily diurnal cycles and broader temporal scales, it is important to continue to assess the role of temperature fluctuation in vector competence in order to accurately model how dynamic climates will influence transmissibility. Further, despite the general trend of increasing vector competence with increasing temperatures, there has been significant variability in experimental studies. Characterizing this variability, and understanding the underlying mechanistic basis is important, both to develop appropriately nuanced models of WNV under changing climates and to inform the development of novel control tools that could combat expected increases in WNV transmission.

2.1 Population and species-specific relationships between temperature and vector competence

Variability in the relationship between competence and temperature among mosquito species and populations is well documented for WNV. For instance, while most experimental studies have found increased viral replication, decreased EIP and increased overall competence for WNV in *Cx. pipiens* (Dohm et al., 2002; Kilpatrick et al., 2008), a study comparing Dutch and Italian populations found increases in competence from 18 °C to 23 °C in both populations, yet a further increase to 28 °C only enhanced competence in the Italian population (Vogels et al., 2017). In addition, studies with *Cx. pipiens* form *molestus*, an autogenous, subterranean, urban-adapted subspecies, demonstrated decreased WNV competence at 28 °C relative to 23 °C (Vogels et al., 2016). For *Cx. quinquefasciatus*, studies again generally demonstrate increased WNV competence with temperature shifts from 25 °C to 30 °C, although a study with a colonized Florida mosquito population additionally demonstrated decreased dissemination at 28 °C relative to either 25 °C or 30 °C (Anderson et al., 2010; Richards et al., 2009). Studies with *Cx. tarsalis* have been more consistent, but the extent of the relationship between temperature and transmission is population-dependent (Danforth et al., 2015; Reisen et al., 2006). These population-specific differences have been more thoroughly studied for *Aedes* mosquitoes (Ciota et al., 2018; Gloria-Soria et al., 2020; Zouache et al., 2014).

This variability in the magnitude of the effect of temperature on vector competence or, in some cases, deviations from expected directions of thermal responses that are population or species-specific, implies that distinct mosquito genetics, transcriptomics, immunity and/or microbiota can influence vector competence in ways that can either reinforce or supersede the independent influence of temperature on viral replication rate. Understanding these complex relationships is critical to predicting how climate change could have asymmetrical effects on WNV competence that are region-specific.

2.2 Mosquito genetics, immunity and climate

Although studies have confirmed genetic variability among mosquito populations with altered WNV competence, similar studies have not been completed to assess the genetic signatures among *Culex* populations that possess varied competence in response to temperature shifts. Insight can be gained from studies with *Aedes* mosquitoes. A study with geographically distinct *Ae. aegypti* from Argentina demonstrated that the influence of temperature on competence for DENV, ZIKV and chikungunya virus (CHIKV) was highly variable among genetically distinct populations (Ciota et al., 2018). Interestingly, this study demonstrated that temperature increases decreased competence for one population derived from a more temperate region. A number of other studies have identified a genetic component for the influence of temperature on competence of *Aedes* mosquitoes (Gloria-Soria et al., 2017; Lambrechts et al., 2011; Mercier et al., 2022; Zouache et al., 2014). The basis of this variation could be partially attributed to the capacity to mount an RNA interference (RNAi) response, yet this has not been adequately studied in adult mosquitoes under variable temperatures (Gloria-Soria et al., 2017). RNAi is considered the primary immune mechanism mounted in defense of arboviruses in mosquitoes (Blair, 2011), and its role in the *Culex*-WNV system is established (Chotkowski et al., 2008; Paradkar et al., 2014). The Janus kinase-signal transducer and activator of transcription pathway (JAK/STAT) additionally contributes to WNV immunity in the mosquito (Colpitts et al., 2012; Paradkar et al., 2012) in conjunction with the RNAi pathway (Ahlers et al., 2019). Although not well-studied in *Culex*, the Toll and immune deficiency (Imd) pathways likely play a secondary antiviral role (Lee et al., 2019). These and other important regulatory pathways have been shown to be bound to transcriptional responses associated with altered temperatures and competence in *Aedes* mosquitoes (Ferreira et al., 2020; Liu et al., 2022). Variability in the capacity to mount these responses under changing environmental conditions is likely to contribute to population-specific differences in vector competence under climate change (Kang et al., 2018). Future studies that aim to further our understanding of the genetic correlates of population-specific transmissibility with increased temperatures could both improve the resolution of transmission models and inform novel vector control strategies.

2.3 Mosquito microbiome and climate

The influence of mosquito microbial populations on arbovirus susceptibility and transmission is well-established (Dennison et al., 2014; Hegde et al., 2015). In particular, the relationship between the vertically transmitted endosymbiont *Wolbachia pipientis* and arbovirus fitness has been extensively studied (Johnson, 2015). The fact that *Wolbachia* is additionally a reproductive parasite that can quickly drive itself through populations has made it an attractive and effective tool for arbovirus control on broad scales, particularly for *Ae. aegypti* (Dutra et al., 2016; Johnson et al., 2015; Martinez et al., 2014).

Unlike *Ae. aegypti,* the primary vectors of WNV are naturally infected with *Wolbachia*. Although nonnative *Wolbachia* is typically more effective at reducing arbovirus fitness and transmission, native *Wolbachia* levels have been demonstrated to perturb WNV susceptibility, replication and competence. Specifically, studies with both *Cx. quinquefasciatus* and *Cx. pipiens* suggest variability in *Wolbachia* is one factor influencing vector competence (Glaser and Meola, 2010; Micieli and Glaser, 2014). Studies in mosquito cell culture also establish a generic capacity of *Wolbachia* to inhibit WNV replication (Ekwudu et al., 2020; Hussain et al., 2013). Interestingly, experiments with *Cx. tarsalis* suggest *Wolbachia* could enhance WNV infection in this species, demonstrating the complexity and specificity of these interactions (Dodson et al., 2014). In addition to *Wolbachia*, generic relationships between microbial diversity and WNV susceptibility have been measured in *Culex* mosquitoes (Zink et al., 2015). Other specific bacterial genera, including *Elizabethkingia* (Onyango et al., 2020a,b) have been implicated in arbovirus competence in other systems, and it is likely that complex interactions between microbial communities, vectors and virus contribute to WNV transmissibility.

Because these interactions are to some extent governed by environmental factors, the influence of dynamic climates on both extrinsic and intrinsic microbial communities could be highly significant in governing prevalence and transmissibility of arboviruses and other vector-borne pathogens. Recent studies with *Aedes* mosquitoes demonstrate that temperature variations in part define gut microbial communities, and that these shifts in microbial signatures contribute to vectoral capacity for arboviruses (Onyango et al., 2020a,b). Prolonged heat stress, in particular, has been shown to decrease *Wolbachia* density (Ulrich et al., 2016). While similar controlled experimental studies have not been completed with *Culex* mosquitoes, these relationships are also likely to be highly relevant to how WNV interacts with its primary vectors. In addition, as with most experimental studies assessing vector competence, previous work focuses exclusively on adult mosquitoes, yet climate variability has profound effects on microbial communities of the aquatic habitats of immature stages of mosquito development. How this variability will influence adult microbiomes, and subsequently competence for arboviruses, is largely unknown. A previous study characterized microbial communities of mosquito species derived from distinct regions of Ontario, Canada over 3 years, and assessed the impact of climate and the relationship to WNV prevalence (Novakova et al., 2017). Microbial signatures were highly distinct among species but also fluctuated significantly within individual seasons. Importantly, shifts in microbiota in *Culex* mosquitoes correlated with fluctuations in WNV abundance and with changes in temperature, but not precipitation. Specifically, increased temperatures were associated with decreased *Wolbachia* levels, potentially contributing to increased transmissibility in subsequent weeks. These findings offer an additional mechanism by which climate change could indirectly contribute to vector competence for WNV and further exacerbate increased population level transmission.

2.4 Additional considerations for climate and vector competence

Although experimental studies investigating the relationship between temperature and competence have focused on adult mosquitoes, temperature variability during immature development could additionally influence competence. Specifically, indirect effects of body size (Rueda et al., 1990) or altered gene expression originating in larval stages could have wide-ranging effects in adult mosquitoes (Muturi et al., 2012). A number of studies assessing this relationship have found that rearing of mosquitoes at lower temperatures increases susceptibility to arboviruses, yet this relationship has been identified predominantly for alphaviruses and their vectors (Kay and Jennings, 2002; Kramer et al., 1998; Turell et al., 2020; Westbrook et al., 2010). Interestingly, there is evidence that the mechanistic basis of this relationship is a destabilizing of the RNAi pathway at cooler temperatures (Adelman et al., 2013), yet this has not been assessed with *Culex* mosquitoes. In addition, a study with WNV and *Cx. tarsalis* found no effect of rearing temperature on competence (Dodson et al., 2012), suggesting that these interactions are unlikely to offset increases in WNV transmission associated with rising temperatures.

Increases in drought in some regions could additionally influence vector competence. Drought is known to be associated with egg retention in *Culex* mosquitoes (Day and Curtis, 1989) and egg retention has been shown to be associated with increase competence for WNV by *Culex quinquefasciatus* (Smartt et al., 2010).

Further, while focus has appropriately been on the primary vectors, secondary vectors are an additional consideration, particularly if they are potential bridge vectors that could facilitate human infection and disease. WNV has been isolated from hundreds of species of mosquitoes worldwide and most species evaluated in the laboratory have been demonstrated to be at least moderately competent (Ciota, 2017). If, on average, competence increases among secondary vectors under climate change, this is likely to have additional consequences for global transmissibility and disease.

Importantly, vector competence is just one component of vectorial capacity, the overall transmission potential of a given mosquito population for a pathogen. Population size, longevity and blood feeding behavior together play a disproportionally significant role in vectorial capacity. Because of this, relatively incompetent mosquitoes can sustain or facilitate arbovirus outbreaks or, conversely, highly competent mosquitoes without frequent host contact or sufficient fitness can at times be inadequate arbovirus vectors. The most significant effects of climate variability are likely to be on mosquito physiology and, subsequently distribution and life history traits. Understanding these relationships is therefore critical to dissecting the broader impact of climate change on WNV transmission.

3. Mosquito physiology and climate

Culex spp. mosquitoes undergo complete metamorphosis, with a life cycle that consists of four stages: egg, larva, pupa and adult. Larval stages can be further delineated into four distinct instars. Generally, this cycle (from egg to egg) takes 2–3 weeks (Crans, 2004). The immature stages of mosquito development occur in dynamic aquatic environments that are significantly influenced by short and long-term environmental and ecological changes. Adults maintain themselves on flowering plants and females mate within days of emergence. Protein acquired through blood meals is required for proper egg development and *Culex* largely prefer avian hosts for blood feeding (Tempelis, 1975). Days after feeding on a vertebrate host, females search for organically rich water sources to lay egg rafts, which are generally laid in clusters of 100–300 eggs (Vinogradova, 2000). All of the biological markers of mosquito fitness, referred to as life history traits, can be greatly influenced by fluctuations in climate, and alterations to these traits have significant implications for vector-borne pathogen transmission.

Mosquito development during the aquatic life phase is accelerated at increased temperature (Alto and Bettinardi, 2013; Ciota et al., 2014; Dodson et al., 2011; Grech et al., 2015; Gunay et al., 2010; Loetti et al., 2011a, 2011b; Mpho et al., 2002; Shelton, 1973; Spanoudis et al., 2018). For *Culex* mosquitoes, experimental studies have confirmed this relationship, with development time of *Cx. pipiens*, *Cx. restuans* and *Cx. quinquefasciatus* inversely correlated to mean temperatures from 16 °C to 30 °C (Ciota et al., 2014). Temperature fluctuations, as well as changes to mean temperature, influence development rates and, like competence, are generally in agreement with mean temperature studies (Spanoudis et al., 2018). The consequence of decreased development time is presumably increased population size, but increased temperature can also lead to increased immature and adult mortality, altered blood feeding, differential mating success, and altered hatch rates (Ciota et al., 2014; Spanoudis et al., 2018). A study with *Cx. quinquefasciatus* in Argentina revealed immature survival was positively correlated with temperatures ranging from 16.6–25.2 °C (Grech et al., 2015). In contrast, Egyptian *Cx. pipiens* were found to have decreased longevity at simulated summer-autumn temperature in comparison to winter-spring temperatures (Abouzied, 2017). Generally speaking, the effect of temperature on immature survival is thought to be unimodal with the optimal temperature for *Cx. quinquefasciatus* 23 °C, *Cx. pipiens* form *molestus* 25 °C, and *Cx. eduardoi* 28 °C (Gunay et al., 2010; V Loetti et al., 2011a; Loetti et al., 2011b; Spanoudis et al., 2018). Body size is inversely correlated with immature development so as temperature increases body size generally decreases (Ciota et al., 2014). Studies with *Aedes* spp. suggest that smaller mosquitoes are more susceptible to certain flaviviruses (Grimstad and Haramis, 1984; Paulson and Hawley, 1991), although there is limited evidence supporting this relationship for *Culex* and WNV (Dodson et al., 2012). Further, a significant correlation between mosquito fitness and body size has not been identified with *Culex spp*.

Findings from studies evaluating the effects of temperature on adult longevity and survivorship together indicate decreased longevity at increased temperatures between 15 °C and 32 °C (Abouzied, 2017; Alto et al., 2014; Andreadis et al., 2014; Ciota et al., 2014; Gunay et al., 2010; Oda et al., 1999; Spanoudis et al., 2018). The thermal limit for the *Culex* genus has been estimated to be 34 °C (Dodson et al., 2011; Loetti et al., 2011a,b; Rueda et al., 1990; Spanoudis et al., 2018). In addition, because adult survival is among the primary factors determining the number of susceptible vectors at a given time, it is also key to determining the likelihood that a mosquito will survive the pathogen EIP. For these reasons, the negative correlation generally observed between temperature and *Culex* longevity to some extent counters increases in WNV transmissibility at higher temperatures resulting from increased development rates and vector competence.

Temperature can additionally influence mosquito host seeking behavior and, subsequently, transmission of WNV. Experimentally, *Cx. quinquefasciatus* and *Cx. pipiens* have been shown to blood feed more frequently as temperature increases, but this was only investigated from 10 °C to 28 °C (Eldridge, 1968; Ruybal et al., 2016), and temperatures above 28 °C could decrease feeding frequency (Ciota et al., 2014). A study using a field acquired *Cx. pipiens* population demonstrated that the cumulative fraction of females taking a second blood meal increased linearly with temperature and the number of days between blood meals (Ruybal et al., 2016). The effect of temperature on fecundity seems to also follow a unimodal response, with an optimal temperature of 24 °C for *Culex. spp.* oviposition (Ciota et al., 2014). Previous studies have shown that increased temperature can reduce *Cx. pipiens* form *molestus* mating activity, resulting in low ovipositing and hatch rates (Oda et al., 1980). Since the primary vectors of WNV in temperate regions overwinter as diapausing adults, and the physiological state of diapause is initiated and terminated by hormonal changes linked to extrinsic photoperiod and temperature, prolonged temperature changes are likely to have a significant effect on early and late season populations (Ciota et al., 2011b; Eldridge, 1968).

In addition to temperature, patterns of precipitation, humidity and drought also have significant effects on *Culex* life history traits and, subsequently WNV transmission. Since these environmental changes are more difficult to simulate in the laboratory, the correlates of fluctuations have been more readily evaluated by modeling relationships between precipitation, *Culex* abundance and WNV (see epidemiological models of West Nile virus and climate section). What is known is that precipitation provides diverse breeding habitats for *Culex* mosquitoes, yet there is a limit to the increase in reproductive output as frequent flooding can be detrimental to hatch rates and immature survival (Watanabe et al., 2017). Conversely, a paucity of oviposition sites can lead to crowding and nutritional stress for larvae. While this has not been shown to effect competence in the lab, crowding is natural setting could confer physiological changes important in transmission (Dodson et al., 2011). Simulated drought in the laboratory, although associated with decreased survival, has been shown to increase blood feeding and overall vectorial capacity (Holmes et al., 2022). Climate can additionally have significant impacts on the salinity, acidity and nutrient contents of aquatic larval habitats, which can indirectly influence mosquito physiology and pathogen transmission (Patrick and Bradley, 2000). Lastly, alterations in levels of ultraviolet radiation and shade which are associated with climate variability could further impact *Culex* fitness (Villena et al., 2018).

3.1 Additional considerations with climate and *Culex* fitness

There are still numerous questions regarding the effects of climate on mosquito physiology and ecology. Precise environmental conditions reflective of natural habitats are unattainable in a laboratory setting and many studies have been completed with colonized mosquitoes or a highly limited number of field populations. Further, most studies have utilized static temperatures rather than attempting to simulate current and predicted future daily fluctuations. In addition, while WNV infection is known to alter *Culex* life history traits including adult survival, blood feeding and fecundity in a species-specific manner, how temperature might impact these relationships remains unstudied (Ciota et al., 2011b; Styer et al., 2007).

For studies that have assessed population and species-specific effects in *Culex* mosquitoes, as with vector competence, distinct associations between environment and life history traits have been identified (Ciota et al., 2014). Similar to competence, population-specific variability can be explained by numerous factors, including mosquito genetics and microbiota. Although studied primarily in *Aedes* and *Anopheles* mosquitoes, the influence and, in fact, requirement of microbiota in mosquito development, adult survival and fecundity is well-established (Coon et al., 2014, 2016, 2022; Valzania et al., 2018). Given the temporal and geographic distinction in *Culex* microbiota (Didion et al., 2021; Duguma et al., 2017; Novakova et al., 2017; Schrieke et al., 2021), the impact of climate on microbial signatures (Novakova et al., 2017) and the previously discussed importance in vector competence, defining how climate change shapes microbial communities of WNV vectors is critical to our understanding of WNV transmissibility.

3.2 Trait-based models for West Nile virus and temperature

Despite the fact that there are still unknowns, the availability of data detailing the relationships between temperature and *Culex* abundance, biting rate, longevity, WNV extrinsic incubation period, and vector competence (Ciota et al., 2014; Dohm et al., 2002; Marini et al., 2016; Ruybal et al., 2016) allow for the development and testing of trait-based models of transmission under temperature change. The basic reproduction number (R_0) is a well-established metric to measure transmissibility and R_0 models for mosquito-borne pathogens were originally developed for malaria (Macdonald, 1961). These models, which are basically an extension of vectorial capacity with additional host factors, include parameters necessary for vector-borne pathogen transmission: adult mosquito mortality rate, biting rate, pathogen development rate, vector competence, infection efficiency, transmission efficiency, mosquito density, host density, and recovery rate (Dietz, 1993). The majority of these variables correspond to mosquito life history traits. As detailed, previous work has suggested that mosquito life history traits often have a unimodal response to temperature, with a temperature minimum, optimum and maximum (Mordecai et al., 2019). Investigators have expanded previous trait-based models to determine temperature-dependent transmission of WNV across *Culex* spp. mosquitoes (Mordecai et al., 2019; Shocket et al., 2020). Shocket et al. compiled published data, trait-based curves and a modified R_0 equation that includes temperature-dependent parameters. Peak transmission of WNV was estimated to occur between 23 °C and 26 °C (Mordecai et al., 2019; Shocket et al., 2020). Optimal R_0 was found to be highest for *Cx. quinquefasciatus*, followed by *Cx. pipiens* and *Cx. tarsalis*, respectively, yet all were approximately 25 °C. Larger species-specific differences were estimated for extremes, with WNV transmission by *Cx. tarsalis* more tolerant to colder temperatures and, surprisingly, WNV transmission by *Cx. pipiens* more heat tolerant (Shocket et al., 2020). As these models are only as good as the experimental data used to inform them, there are important caveats to acknowledge. First, temperature is treated as a constant, and it remains unclear if this is reflective of fluctuations in nature. Second, experiments at times utilized colonized mosquito or single populations and are unlikely to have captured the full range of relationships between temperature and *Culex spp*.

Despite these limitations, trait-based models have important utility for predicting changes to WNV transmission under future climate scenarios. A previous study compared a trait-based approach to two random forest models of WNV transmission under climate change in New York and Connecticut (Keyel et al., 2021). The results of the trait-based mosquito model showed a decrease of transmission in current high-risk areas. A similar study modeling WNV transmission in Europe demonstrated wide temperature suitability for the vector *Cx. pipiens* (Pol et al., 2022).

4. Epidemiological models of West Nile virus and climate

4.1 Temperature and West Nile virus

A large effort has been underway for over 50 years to study the effect of temperature on the transmission of vector-borne diseases. Several studies use models that consider historical mosquito sampling, climate, and clinical data to describe WNV risk factors and predictors. Studies have identified that mean, minimum and maximum temperatures can influence WNV transmission. Mosquito abundance has been shown to be positively associated with mean weekly temperature (Moise et al., 2018). Higher daily temperature is associated with increased WNV incidence in Italy (Riccò et al., 2021). Seasonal differences in minimum and maximum temperature also are linked to WNV incidence (Keyel et al., 2019; Marini et al., 2020a,b), yet the effect may not be significant when single day anomalies are considered. A lag effect of temperature has been shown whereby high temperature (greater than 25 °C) during the previous month strongly positively influenced WNV seroprevalence in sentinel chickens (Kernbach et al., 2021), and employing two-to-three-week lags further demonstrated a correlation with increased WNV incidence in both the U.S. and Romania (Cotar et al., 2016; Myer et al., 2017; Myer and Johnston, 2019). Temperature lags also have been examined in the ArboMAP model (Davis et al., 2017, 2018), and in a functional linear modeling approach (Smith et al., 2020).

The role of ambient temperature has been well studied, and a 1 °C increase over a 5-week span is associated with reduced *Culex* spp. larval habitats in non-residential areas in Singapore (Soh and Aik, 2021). Higher average temperatures 2–4 weeks prior contributed to a higher probability of human WNV cases in certain areas of Chicago, IL (Karki et al., 2020). In Greece, increases in temperature were correlated with increased human WNV cases based on retrospective models (Stilianakis et al., 2016). A large proportion of studies focus on ambient temperature, but soil temperature can also be a predictor of WNV outbreaks (Stilianakis et al., 2016). Further, although average temperature is important, temperature anomalies during distinct seasons can have differential effects.

4.1.1 Winter

Although there is a low risk of WNV transmission during the winter months, studies have found that climatic patterns during cold months can influence WNV incidence during transmission season. Warmer temperatures in the winter are associated with increased WNV transmission in the continental U.S., Europe, and Russia (Hahn et al., 2015; Karki et al., 2020; Manore et al., 2014; Mihailović et al., 2020; Platonov et al., 2008, 2014). It is estimated that virus development is unlikely to occur below 14.3 °C in *Cx. tarsalis* and such observations support the hypothesis that winter temperatures reduce WNV replication (Reisen et al., 2006).

These studies have important implications for WNV overwintering. Current models suggest that winter crow roosts could allow for WNV persistence through winter temperatures (Montecino-Latorre and Barker, 2018) and infected crows have been found in the middle of winter

(Dawson et al., 2007). Northern Cardinals with prior WNV infections had increased winter mortality (Ward et al., 2010), and therefore harsher winters could decrease the likelihood of actively infected birds surviving to pass on the infection. However, it is generally accepted that in temperate regions WNV has the capacity to overwinter in diapausing adult mosquitoes. While cold temperatures may impede replication, WNV amplification could recommence following springtime emergence. Indeed, WNV infected *Culex* have been collected from hibernacula during the winter months (Farajollahi et al., 2005; Kampen et al., 2021; Nasci et al., 2001; Rudolf et al., 2017), commencement of WNV activity is generally correlated to *Culex* emergence (Ciota et al., 2011a), and genetic studies suggest local or regional maintenance of WNV in temperate areas (Bertolotti et al., 2008; Ehrbar et al., 2017). Given the various hosts, vectors and landscapes that shape WNV transmission cycles in distinct regions, it has been suggested that several pathways may exist for WNV overwintering (Reisen and Wheeler, 2019).

4.1.2 Spring

Spring temperatures shape the mosquito breeding season, which can help amplify WNV during the early transmission season (Marini et al., 2020a,b). Warm and dry conditions in early spring have been associated with increased annual WNV infection rates in *Culex* spp. mosquitoes (Little et al., 2016). Increased maximum temperature in the spring is also correlated with early season amplification of WNV (Marini et al., 2020a,b). Higher temperatures in late spring, May–June, have been related to WNV transmission in Russia (Platonov et al., 2014). Temperature anomalies during the month of May contributed to increases in the total number of human WNV cases in Greece (Angelou et al., 2021). County level models utilizing retrospective human WNV cases and meteorological variables have been successful in predicting an earlier-than-normal start to the WNV season during the 2016 outbreak in South Dakota (Davis et al., 2017). In Italy, warmer temperatures in the spring were associated with earlier start to transmission season and increased overall length of the season (Rosà et al., 2014).

4.1.3 Summer

Summer temperatures, specifically above average monthly temperatures, were linked to WNV outbreaks from 2002 to 2004 in the continental U.S. (Reisen et al., 2006). Additionally, increased temperatures during the summer months were historically associated with increased WNV cases in Russia and Europe, as well as recent outbreaks in Europe (Marcantonio et al., 2015; Paz and Semenza, 2013; Platonov et al., 2014; Tran et al., 2014). The primary WNV vector species are active at dusk and night, and often take refuge during the day (Danforth et al., 2016). As a consequence, mean minimum (i.e., nighttime) temperatures may be more strongly associated with WNV risk than diurnal temperatures (maximums). Concordantly, mean minimum temperature during the summer months of July–September has been shown to be a strong predictor of human WNV cases in a temperate region (Keyel et al., 2019).

4.1.4 Fall

Cooler temperatures in the fall can result in an early conclusion of the WNV transmission season. In contrast, increases in average temperature across Italy presented favorable conditions for WNV transmission which resulted in an extended transmission season (Candeloro et al., 2020). Warmer fall temperatures may extend the mosquito season, but generally have not been identified as a significant predictor of WNV transmission (Ciota and Keyel, 2019). While cooler temperatures in the fall will influence mosquito survival and diapause initiation, changes in host behavior and abundance independent of temperature may be the primary drivers of transmission in the fall (Kilpatrick et al., 2006).

4.2 The influence of precipitation, humidity, and soil moisture on West Nile virus

The influence of precipitation on the incidence of WNV is well studied. The general trend is that lower precipitation is associated with increased mosquito infections and human incidence, yet this relationship is complex and contingent on numerous factors (Aharonson-Raz et al., 2014; Ruiz et al., 2010). For example, a study investigating climate during an early outbreak of WNV in the U.S. conversely found heavy precipitation increased human WNV cases (Soverow et al., 2009). Patterns of precipitation, rather than abundance alone, are clearly important. For instance, drier conditions followed by wetter periods have been linked to increased WNV incidence in Illinois and Nebraska (Ruiz et al., 2010; Smith et al., 2020). Heavy rainfall can be negatively correlated with *Culex* spp. larval habitat abundance (Gardner et al., 2012; Jones et al., 2012; Soh and Aik, 2021), with longer exposure to rainfall shown to flush out *Cx. pipiens* larvae (Koenraadt and Harrington, 2008). In contrast, early precipitation contributed to increased

Cx. pipiens abundance in Italy (Rosà et al., 2014), and increased precipitation for up to 35 days in Canada was found to accompany increased *Cx. pipiens* and *Cx. restuans* abundance, demonstrating a threshold for the influence of precipitation in temperate climates (Wang et al., 2011). Likely because of this complexity, some studies have indicated no correlation between rainfall and mosquito abundance (Damos and Caballero, 2021).

Seasonal variation in precipitation patterns is clearly important in driving WNV activity. Wet winters in New York State are correlated with increased WNV prevalence in *Culex* spp. mosquitoes (Shaman et al., 2011). Several studies have investigated spring precipitation and its influence on WNV transmission. There is an association between early spring precipitation and increased standing water (Marcantonio et al., 2015) and studies in Europe show a positive correlation between WNV fever incidence, the total number of days with precipitation in late winter-spring and irrigated croplands. This is likely because standing water is required for larval development of mosquitoes, and this is particularly important for early season amplification. In agreement with this, studies in Texas and New York indicate wet springs with dry summers are linked to increased WNV transmission (Shaman et al., 2011; Ukawuba and Shaman, 2018). The contribution of precipitation to WNV incidence is likely variable by region. In the eastern U.S. outbreaks of WNV have been found to be preceded by above average rainfall, while in the western U.S. below average rainfall preceded outbreaks (Landesman et al., 2007). *Culex* spp. mosquitoes vary in their distribution (Fig. 1), with urban, container breeding *Cx. pipiens* and *Cx. quinquefasciatus* in the eastern U.S. and the agrarian *Cx. tarsalis* in the western U.S. These regional differences result in distinct habitat preferences and relationships between precipitation and population growth. Lastly, the length of the season has been associated with precipitation, with early spring rainfall shown to shorten the transmission season and late season rainfall to extend the transmission season (Rosà et al., 2014).

Humidity is positively associated with adult *Culex* spp. activity (Seah et al., 2021) and WNV cases in Italy, Madagascar, and the U.S. (Olive et al., 2021; Riccò et al., 2021; Soverow et al., 2009). This linear effect of relative humidity and WNV incidence is not found in all regions; studies in Israel and Europe show humidity may not be the best predictor of WNV incidence (Lourenço et al., 2020; Paz and Semenza, 2013). In addition, low relative humidity was linked to a higher number of human WNV cases in Greece (Stilianakis et al., 2016). Recent trends in precipitation and humidity may be best indicated by measures of soil moisture. Indeed, studies in Texas found that soil moisture during the summer was the strongest predictor of WNV incidence (Ukawuba and Shaman, 2018). Data suggests WNV transmission demonstrates a unimodal response to soil moisture with higher transmission associated with both high and low moisture levels (Keyel et al., 2019). The relationship between WNV and moisture variables may also be non-linear and may depend on other covariates (Keyel et al., 2019).

In particular, the interaction of temperature and precipitation is important. Drought can lead to the consolidation and confinement of both birds and mosquitoes near aquatic sources (Shaman et al., 2005). The lack of dispersal could to some extent facilitate bird to bird WNV transmission, which has been observed under experimental conditions (Hartemink et al., 2007; Komar et al., 2003) and, more importantly, increase contact between birds and mosquitoes to facilitate subsequent amplification and spread (Shaman et al., 2005). In support of this, numerous studies have shown that drought can lead to increased WNV risk (Epstein and Defilippo, 2001; Johnson and Sukhdeo, 2013; Little et al., 2016; Wang et al., 2010).

4.3 Vector and host distribution

Temperature has been shown to influence the distribution of vectors and hosts (Lebl et al., 2013). To date, *Culex* spp. have been identified on every continent, excluding Antarctica (Farajollahi et al., 2011; Gorris et al., 2021; Wu et al., 2019) (Fig. 1). Amplifying avian hosts of WNV have additionally been identified across the globe. Over 330 avian species have been infected with WNV to date (CDC, 2017). Competent non-avian hosts have the potential to be important in regional dynamics (e.g., alligators (Klenk et al., 2004) and Fox squirrels (Root et al., 2006)). Altered distributions of both mosquitoes and birds could facilitate the expansion of WNV into regions with additional suitable hosts, which could result in increased WNV transmission. As mosquito and avian hosts play the primary role in WNV transmission factors that influence their distribution are reviewed below.

Species distribution modeling (SDM) has been used to determine vector distribution. The scope of such models is variable, *Culex* spp. distribution has been modeled across continents showing habitat suitability (Alaniz et al., 2018; Conte et al., 2015; Farajollahi et al., 2011; Hongoh et al., 2012; Samy et al., 2016). The effects of climate change, linked to increased greenhouse gas emissions, are predicted to shift mosquito abundance (Morin and Comrie, 2010) and distribution (Samy et al., 2016). *Cx. pipiens* are predicted to become more widely present in Canada (Chen et al., 2013; Hongoh et al., 2012), with northward shifts across the northern hemisphere

(Hoover and Barker, 2016). Current research suggests decreases in *Cx. quinquefasciatus* in California and Florida during the primary transmission season but this will likely result in increased mosquito abundance during the winter months (Morin and Comrie, 2010; Samy et al., 2016).

Non-climatic factors also influence species distribution, and these may modify or obscure climate-WNV relationships. A study in the San Joaquin Valley of California found areas used for agriculture generally have relatively lower temperature and biting rates, resulting in reduced probability of transmission (Boser et al., 2021). This may also be due to reduced human populations in these areas, as total human population is positively associated with probability of WNV cases and mosquito abundance (Andreadis et al., 2004; Karki et al., 2020), although some of the highest incidence areas for WNV in the U.S. are located in low-population areas (CDC, 2020). As urbanization increases across the globe, the incidence of artificial light at night (ALAN) increases. Studies show that ALAN may contribute to spatiotemporal changes in WNV risk (Kernbach et al., 2021; Marzluff, 2001). Higher population and higher proportion of urban light intensity has been associated with increased WNV cases (Karki et al., 2020). Human population density and landcover were found to influence *Cx. pipiens* distribution (Conley et al., 2014; Rochlin et al., 2008). In Portugal, croplands are linked to increased reports of WNV (Lourenço et al., 2022). In Italy, proximity to rice fields was related to increased vector abundance but was not a significant predictor of *Culex* spp. abundance (Rosà et al., 2014). Land cover variables such as proportion of open water, grassland, and deciduous forests are connected to reduced risk of WNV in the Chicago area (Karki et al., 2020). Altitude and distance to water both have been shown to play a role in WNV cases in Greece and Russia (Shartova et al., 2022; Valiakos et al., 2014). In New York State, wetlands and areas closer to the shore contributed to a lower incidence of WNV (Myer et al., 2017; Myer and Johnston, 2019). A retrospective study of WNV lineage dispersal found WNV is generally dispersed in areas of urban coverage or shrublands (Dellicour et al., 2020).

Climate change and globalization will likely alter avian distributions (Carroll et al., 2015; Huntley et al., 2006). Predictions of shifts in avian species richness under the representative concentration pathway 4.5, which is a greenhouse gas concentration trajectory adopted by the IPCC, suggest that the Americas will likely see a decline in the richness of species of conservation concern, with the largest declines in the southern Amazon region (Voskamp et al., 2021). Climate change has significant impacts on avian diversity, patterns of dispersal, behavior, physiology and fitness, all of which can have profound effects on WNV amplification and patterns of transmission (Voskamp et al., 2021). WNV has been isolated from hundreds of avian species, yet host competence is highly variable (Kilpatrick et al., 2006; Komar et al., 2003; Pérez-Ramírez et al., 2014; Reisen et al., 2005).

For instance, corvids (crows, jays, magpies) are known to support high viremia levels, yet also be highly susceptible to disease (Weingartl et al., 2004). On the other hand, avian hosts such as galliforms (waterfowl) and columbiformes (doves) have low WNV competence. Passeriformes with moderate to high viremia levels that are not generally subject to high levels of virulence (for example, house sparrows and American robins) are the most ideal WNV amplifying hosts. In addition, important host feeding preferences for these species have been shown for *Culex* mosquitoes (Hamer et al., 2009; Kilpatrick et al., 2006; Komar et al., 2013). For these reasons, changes to host distribution or species richness as a result of climate change could facilitate large shifts in population level transmissibility of WNV. The relationship between species diversity and WNV transmission is highly contingent on regional variability and seasonality (Kain and Bolker, 2019). If decreased richness equates to higher proportions of competent hosts, this could increase amplification. On the other hand, if there are fewer competent hosts this could dampen amplification.

Climate change effects on avian phenology may also affect WNV dynamics. The timing of migration and/or duration of the breeding season may shift. Studies have shown timing of nesting relative to mosquito emergence can influence amplification dynamics (Caillout et al., 2013; Robertson and Caillouët, 2016). While breeding, birds remain locally in the vicinity of a nest. Once young are fledged, birds can move to locations with fewer mosquitoes. Such changes in breeding behavior have been identified in American Robins and the proportion of mosquitoes feeding on mammalian hosts has been shown to be indirectly correlated to the presence of robin populations (Kilpatrick et al., 2006). Thus, changes to the timing of breeding due to climate change may affect spillover to humans. Migrating birds are thought to play a major role in the amplification and transmission of WNV (Owen et al., 2006). Indeed, migratory birds are known to introduce WNV to non-endemic areas (Jourdain et al., 2011) and shifts in climate suitability could lead to future amplification in these locations.

Finally, climate change could affect avian physiology through direct thermal/water stresses, or indirectly through changes to land use, habitat fragmentation, or nutritional availability. Food stress in American Robins was shown to increase WNV viremia, leading to substantial increases in the risk of vector-borne disease transmission (Owen et al., 2021).

5. The influence of temperature on West Nile virus diversity and evolution

WNV and other arboviruses are almost exclusively RNA viruses that lack proofreading mechanisms, exist within hosts and vectors as highly diverse populations, and have the capacity for rapid adaptation and diversification. While studies assessing how temperature could influence fidelity (the intrinsic error rate of genome replication) are lacking, increased replication rates resulting from increased temperatures should by definition increase mutation frequency and, subsequently, diversification and evolutionary rates (Fig. 4). Further, as WNV temperature sensitivity itself has a genetic basis, strains with increased fitness at higher temperatures would be predicted to emerge under sustained warming. In fact, it could be argued that adaptive trade-offs are to some extent driven by host-specific temperature requirements of WNV and other arboviruses (i.e., replication in ectothermic mosquitoes vs. warm-blooded vertebrates), and that as mean temperatures rise these tradeoffs could be diminished, further accelerating the emergence of higher fitness strains. A previous study assessing the influence of temperature on WNV evolution and adaptation in mosquito cell

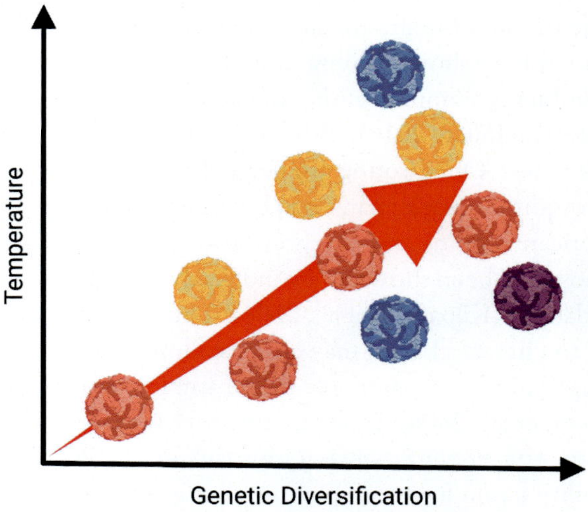

Fig. 4 The relationship between temperature and virus diversification. Increased temperatures generally accelerate viral replication rates in ectothermic vectors and facilitate virus dispersal. This increases genetic and phenotypic diversification and could facilitate the emergence of novel West Nile virus strains with altered transmissibility and virulence.

culture compared experimental passage at 30 °C to passage 25 °C and demonstrated increased genetic diversity, particularly on the amino acid level, and an increased probability of emergence of broadly adaptive strains with passage at the higher temperature (Fay et al., 2021). Whether or not selective pressures on WNV are variable under distinct diurnal temperature fluctuations in *Culex* mosquitoes has not been adequately assessed, yet studies with ZIKV and *Ae. aegypti* suggest broad fluctuations could decrease the likelihood of positive selection (Murrieta et al., 2021). Given that measurable temperature increases have already been documented over the known evolutionary history of WNV, the effect of temperature on WNV evolution and adaptation can to some extent be assessed with historical data. Indeed, a study assessing the relationship between WNV dispersal and numerous factors, found that the only statistically significant environmental influence associated with increased genetic diversification in nature is temperature, with elevated temperature driving the dispersal and emergence of novel strains (Dellicour et al., 2020). Further, the WN02 genotype, which rapidly displaced the NY99 genotype in the Americas and was experimentally demonstrated to be more transmissible by *Culex* mosquitoes, was additionally shown to have an increased advantage at higher temperatures relative to historic strains. While the role of temperature in driving the emergence of novel high fitness strains has not been thoroughly assessed, new genotypes with positively selected substitutions have emerged since the WN02 (Bialosuknia et al., 2019; Ebel et al., 2004; McMullen et al., 2011), including the recent identification of the NY10 genotype in New York State, which has largely displaced previous genotypes and is associated with both increased WNV prevalence in mosquitoes and increased transmissibility in the lab (Bialosuknia et al., 2022). These data together support the idea that rising temperatures are increasing both genetic and phenotypic diversity and accelerating the likelihood of emergence of strains with increased transmissibility (Fig. 4).

While this phenomenon could further exacerbate increased prevalence under climate change it is again important to consider the complexity and specificity of vector-virus-environment interactions. Particularly, there may be virulence trade-offs that counter the selective advantage of high fitness strains. For instance, a previous experimental study demonstrated that a mosquito-adapted WNV strain selected for increased competence in *Cx. pipiens* also decreased mosquito fitness and overall vectorial capacity (Ciota et al., 2012). Lastly, as WNV encounters and infects a wide breadth of hosts and vectors and has been shown to have the capacity to utilize unique systems in natural and experimental settings, the emergence of novel strains could have additional consequences outside of the primary enzootic cycle of WNV that could impact distribution and prevalence in the future.

6. Concluding remarks

Climate change has already had a significant impact on the prevalence and distribution of WNV and other vector-borne pathogens, and the magnitude and pace of these changes is likely to grow without innovative vector control or significant curbing of anthropogenic contributors to global warming. The challenge to understanding the trajectory of WNV transmission under future climate change is that it requires us to fully consider not just how changing environments influence current biological systems (Fig. 5), but also how these systems will themselves evolve and interact with future climates.

While it is important to both define the historic relationship between incidence and environmental fluctuation and to simulate future conditions

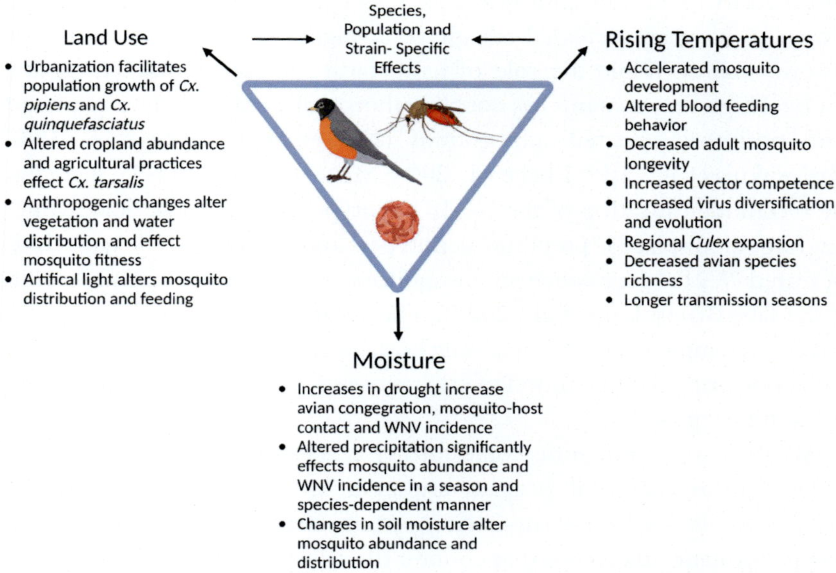

Fig. 5 The complex relationship between climate and West Nile virus (WNV). Increases in temperature, variability in moisture and changes to land use occurring concurrently with climate change impact the vector, virus and host in numerous direct and indirect ways. While general trends have emerged that inform relationships between climate and WNV incidence, these effects are regionally distinct because of unique landscapes, distinct vector and host distributions, population-specific effects and variability in WNV strain.

with current populations in the laboratory, such studies do not fully reflect interactions in dynamic systems subject to evolutionary forces. In addition to not knowing the fitness and virulence of future viral strains, or the precise composition and consequences of future microbial communities, host and vectors are themselves subject to evolutionary pressures. There is a general lack of understanding regarding the specific genetic and physiological correlates of climatic adaptation among mosquitoes, as well as the extent to which adaptive evolution could alter the relationships between mosquito life history traits and climate through time. Important studies have begun to reveal the correlates and evolvability of thermal tolerance in *Drosophila* (Rodrigues et al., 2022; Schou et al., 2014). Similar studies identifying genetic signatures in natural and experimentally evolved populations of mosquitoes could begin to shed some light on the potential changes to mosquito population structure and vectorial capacity under climate change. While the pace and extent of evolution among avian populations may not be equivalent to that of vector populations, and certainly not to that of a RNA virus, changes to avian physiology and behavior resulting from climate adaptation could also have consequences for pathogen transmission (Bonamour et al., 2019).

It is also important to differentiate transmission and prevalence from human disease. While historical mosquito surveillance efforts predictably demonstrate a correlation between WNV prevalence in *Culex* spp. mosquitoes and human cases, the magnitude of this relationship is variable (Bialosuknia et al., 2022). This is partly attributed to geographical or temporal variability in testing and reporting, or to the inherent stochasticity of a relatively rare neurological disease manifestation, but there is also likely important variability in strain virulence and spillover frequency. Climate change could have a significant influence on both of these factors. Virulence is not a trait that is generally independently selected for but rather one that evolves as a byproduct of maximizing transmission (Alizon et al., 2009; Lipsitch and Moxon, 1997). This makes virulence evolution difficult to predict, and this is particularly true for enzootic pathogens like WNV for which no perpetual selective pressure is exerted by humans. The likelihood of spillover could also be dramatically impacted by alterations to host and mosquito populations, which could result from measurable declines in avian populations from widespread WNV infection (Kilpatrick and Wheeler, 2019) and from alterations in feeding behavior of primary vectors and/or secondary bridge vectors (Kilpatrick et al., 2005, 2006; Levine et al., 2016, 2017).

Lastly, the influence of climate on WNV transmission and disease is not occurring independent of epidemiological shifts in prevalence and distribution of other mosquito-borne pathogens. There has been a marked increase in both incidence and geographic range of mosquito-borne viruses and their vectors in recent decades, and this has resulted in the co-occurrence of numerous arboviruses in many regions throughout the globe (Ciota, 2019). The potential direct and indirect interactions among distinct and overlapping virus-vector-host systems should be more thoroughly considered in future studies. While the complexity and uncertainty inherent to vector-borne disease transmission is daunting, significant advancements in our understanding of the influence of environmental fluctuation on these transmission networks have been made in recent years (Bellone and Failloux, 2020; Franklinos et al., 2019; Mordecai et al., 2019; Rocklöv and Dubrow, 2020; Sadeghieh et al., 2020). As ongoing research will continue to unravel the multifaceted implications of climate change on vector-borne pathogens, informed, innovative public health initiatives should be prioritized to combat inevitable shifts in disease transmission globally.

References

Abouzied, E.M., 2017. Life table analysis of culex pipiens under simulated weather conditions in Egypt. J Am Mosq. Contr 33, 16–24. https://doi.org/10.2987/16-6608.1.

Adelman, Z.N., Anderson, M.A.E., Wiley, M.R., Murreddu, M.G., Samuel, G.H., Morazzani, E.M., Myles, K.M., 2013. Cooler temperatures destabilize RNA interference and increase susceptibility of disease vector mosquitoes to viral infection. Plos Neglect. Trop. D 7, e2239. https://doi.org/10.1371/journal.pntd.0002239.

Aharonson-Raz, K., Lichter-Peled, A., Tal, S., Gelman, B., Cohen, D., Klement, E., Steinman, A., 2014. Spatial and temporal distribution of West Nile virus in horses in Israel (1997–2013) - from endemic to epidemics. Plos One 9, e113149. https://doi.org/10.1371/journal.pone.0113149.

Ahlers, L.R.H., Trammell, C.E., Carrell, G.F., Mackinnon, S., Torrevillas, B.K., Chow, C.Y., Luckhart, S., Goodman, A.G., 2019. Insulin potentiates JAK/STAT signaling to broadly inhibit flavivirus replication in insect vectors. Cell Rep. 29, 1946–1960. e5. https://doi.org/10.1016/j.celrep.2019.10.029.

Alaniz, A.J., Carvajal, M.A., Bacigalupo, A., Cattan, P.E., 2018. Global spatial assessment of Aedes aegypti and Culex quinquefasciatus: a scenario of zika virus exposure. Epidemiol. Infect. 147, e52. https://doi.org/10.1017/s0950268818003102.

Aliota, M.T., Kramer, L.D., 2012. Replication of West Nile virus, Rabensburg lineage in mammalian cells is restricted by temperature. Parasit Vectors 5, 293. https://doi.org/10.1186/1756-3305-5-293.

Alizon, S., Hurford, A., Mideo, N., Baalen, M.V., 2009. Virulence evolution and the trade-off hypothesis: history, current state of affairs and the future. J. Evol. Biol. 22, 245–259. https://doi.org/10.1111/j.1420-9101.2008.01658.x.

Alto, B.W., Bettinardi, D., 2013. Temperature and dengue virus infection in mosquitoes: independent effects on the immature and adult stages. Am. J. Trop. Med. Hyg. 88, 497–505. https://doi.org/10.4269/ajtmh.12-0421.

Alto, B.W., Richards, S.L., Anderson, S.L., Lord, C.C., 2014. Survival of West Nile virus-challenged southern house mosquitoes, Culex pipiens quinquefasciatus, in relation to environmental temperatures. J. Vector Ecol. 39, 123–133. https://doi.org/10.1111/j.1948-7134.2014.12078.x.

Anderson, S.L., Richards, S.L., Tabachnick, W.J., Smartt, C.T., 2010. Effects of West Nile virus dose and extrinsic incubation temperature on temporal progression of vector competence in Culex pipiens quinquefasciatus. J Am Mosq. Contr 26, 103–107. https://doi.org/10.2987/09-5926.1.

Andrade, C.C., Maharaj, P.D., Reisen, W.K., Brault, A.C., 2011. North American West Nile virus genotype isolates demonstrate differential replicative capacities in response to temperature. J. Gen. Virol. 92, 2523–2533. https://doi.org/10.1099/vir.0.032318-0.

Andreadis, T.G., Anderson, J.F., Vossbrinck, C.R., Main, A.J., 2004. Epidemiology of West Nile virus in Connecticut: a five-year analysis of mosquito data 19992003. Vector-borne Zoonot 4, 360–378. https://doi.org/10.1089/vbz.2004.4.360.

Andreadis, S.S., Dimotsiou, O.C., Savopoulou-Soultani, M., 2014. Variation in adult longevity of Culex pipiens f. pipiens, vector of the West Nile virus. Parasitol. Res. 113, 4315–4319. https://doi.org/10.1007/s00436-014-4152-x.

Angelou, A., Kioutsioukis, I., Stilianakis, N.I., 2021. A climate-dependent spatial epidemiological model for the transmission risk of West Nile virus at local scale. One Heal 100330. https://doi.org/10.1016/j.onehlt.2021.100330.

Bakonyi, T., Ivanics, E., Erdélyi, K., Ursu, K., Ferenczi, E., Weissenböck, H., Nowotny, N., 2006. Lineage 1 and 2 strains of encephalitic West Nile virus, central Europe. Emerg. Infect. Dis. 12 (4), 618–623. https://doi.org/10.3201/eid1204.051379.

Bakonyi, T., Ferenczi, E., Erdélyi, K., Kutasi, O., Csörgő, T., Seidel, B., Weissenböck, H., Brugger, K., Bán, E., Nowotny, N., 2013. Explosive spread of a neuroinvasive lineage 2 West Nile virus in Central Europe, 2008/2009. Vet. Microbiol. 165, 61–70. https://doi.org/10.1016/j.vetmic.2013.03.005.

Bellone, R., Failloux, A.-B., 2020. The role of temperature in shaping mosquito-borne viruses transmission. Front. Microbiol. 11, 584846. https://doi.org/10.3389/fmicb.2020.584846.

Bernkopf, H., Levine, S., Nerson, R., 1953. Isolation of West Nile virus in Israel. J Infect Dis 93, 207–218. https://doi.org/10.1093/infdis/93.3.207.

Bertolotti, L., Kitron, U.D., Walker, E.D., Ruiz, M.O., Brawn, J.D., Loss, S.R., Hamer, G.L., Goldberg, T.L., 2008. Fine-scale genetic variation and evolution of West Nile virus in a transmission "hot spot" in suburban Chicago, USA. Virology 374, 381–389. https://doi.org/10.1016/j.virol.2007.12.040.

Bialosuknia, S.M., Tan, Y., Zink, S.D., Koetzner, C.A., Maffei, J.G., Halpin, R.A., Muller, E., Novatny, M., Shilts, M., Fedorova, N.B., Amedeo, P., Das, S.R., Pickett, B., Kramer, L.D., Ciota, A.T., 2019. Evolutionary dynamics and molecular epidemiology of West Nile virus in New York state: 1999–2015. Virus Evol 5. https://doi.org/10.1093/ve/vez020.

Bialosuknia, S.M., Dupuis, A.P., Zink, S.D., Koetzner, C.A., Maffei, J.G., Owen, J.C., Landwerlen, H., Kramer, L.D., Ciota, A.T., 2022. Adaptive evolution of West Nile virus facilitated increased transmissibility and prevalence in New York state: adaptive evolution of West Nile virus in New York state. Emerg Microbes Infec, 1–32. https://doi.org/10.1080/22221751.2022.2056521.

Blair, C.D., 2011. Mosquito RNAi is the major innate immune pathway controlling arbovirus infection and transmission. Future Microbiol. 6, 265–277. https://doi.org/10.2217/fmb.11.11.

Bonamour, S., Chevin, L.-M., Charmantier, A., Teplitsky, C., 2019. Phenotypic plasticity in response to climate change: the importance of cue variation. Philos. Trans. R. Soc. B 374, 20180178. https://doi.org/10.1098/rstb.2018.0178.

Bondre, V.P., Jadi, R.S., Mishra, A.C., Yergolkar, P.N., Arankalle, V.A., 2007. West Nile virus isolates from India: evidence for a distinct genetic lineage. J. Gen. Virol. 88, 875–884. https://doi.org/10.1099/vir.0.82403-0.

Boser, A., Sousa, D., Larsen, A., MacDonald, A., 2021. Micro-climate to macro-risk: mapping fine scale differences in mosquito-borne disease risk using remote sensing. Environ. Res. Lett. 16, 124014. https://doi.org/10.1088/1748-9326/ac3589.

Brault, A.C., Huang, C.Y.-H., Langevin, S.A., Kinney, R.M., Bowen, R.A., Ramey, W.N., Panella, N.A., Holmes, E.C., Powers, A.M., Miller, B.R., 2007. A single positively selected West Nile viral mutation confers increased virogenesis in American crows. Nat. Genet. 39, 1162–1166. https://doi.org/10.1038/ng2097.

Briese, T., Jia, X.-Y., Huang, C., Grady, L.J., Lipkin, W.L., 1999. Identification of a Kunjin/West Nile-like flavivirus in brains of patients with New York encephalitis. Lancet 354, 1261–1262. https://doi.org/10.1016/s0140-6736(99)04576-6.

Brinton, M.A., 2002. The molecular biology of West Nile virus: a new invader of the Western hemisphere. Annu. Rev. Microbiol. 56, 371–402. https://doi.org/10.1146/annurev.micro.56.012302.160654.

Caillout, K.A., Riggan, A.E., Bulluck, L.P., Carlson, J.C., Sabo, R.T., 2013. Nesting bird host funnel increases mosquito-bird contact rate. J. Med. Entomol. 50, 462–466. https://doi.org/10.1603/me12183.

Caldwell, H.S., Pata, J.D., Ciota, A.T., 2022. The role of the flavivirus replicase in viral diversity and adaptation. Viruses 14, 1076. https://doi.org/10.3390/v14051076.

Candeloro, L., Ippoliti, C., Iapaolo, F., Monaco, F., Morelli, D., Cuccu, R., Fronte, P., Calderara, S., Vincenzi, S., Porrello, A., D'Alterio, N., Calistri, P., Conte, A., 2020. Predicting WNV circulation in Italy using earth observation data and extreme gradient boosting model. Remote Sens. (Basel) 12, 3064. https://doi.org/10.3390/rs12183064.

Carroll, C., Lawler, J.J., Roberts, D.R., Hamann, A., 2015. Biotic and climatic velocity identify contrasting areas of vulnerability to climate change. Plos One 10, e0140486. https://doi.org/10.1371/journal.pone.0140486.

CDC, 2017. Species of dead birds in which West Nile virus has been detected, United States, 1999–2016.

CDC, 2020. Final Cumulative Maps & Data for 1999–2019. Centers for Disease Control and Prevention [WWW Document]. URL https://www.cdc.gov/westnile/statsmaps/cumMapsData.html#one. (accessed 22).

CDC, 2022a. West Nile [WWW Document]. URL https://www.cdc.gov/westnile/statsmaps/index.html (accessed 22).

CDC, 2022b. West Nile Virus Statistics and Map [WWW Document]. 2022. URL https://www.cdc.gov/westnile/statsmaps/index.html (accessed 22).

Chancey, C., Grinev, A., Volkova, E., Rios, M., 2015. The global ecology and epidemiology of West Nile virus. Biomed. Res. Int. 2015, 1–20. https://doi.org/10.1155/2015/376230.

Charrel, R.N., Brault, A.C., Gallian, P., Lemasson, J.-J., Murgue, B., Murri, S., Pastorino, B., Zeller, H., Chesse, R.d., Micco, P.d., Lamballerie, X.d., 2003. Evolutionary relationship between Old World West Nile virus strains evidence for viral gene flow between africa, the middle east, and europe. Virology 315, 381–388. https://doi.org/10.1016/s0042-6822(03)00536-1.

Chen, C.C., Jenkins, E., Epp, T., Waldner, C., Curry, P.S., Soos, C., 2013. Climate change and West Nile virus in a highly endemic region of North America. Int. J. Environ. Res. Pu 10, 3052–3071. https://doi.org/10.3390/ijerph10073052.

Chotkowski, H.L., Ciota, A.T., Jia, Y., Puig-Basagoiti, F., Kramer, L.D., Shi, P.-Y., Glaser, R.L., 2008. West Nile virus infection of Drosophila melanogaster induces a protective RNAi response. Virology 377, 197–206. https://doi.org/10.1016/j.virol.2008.04.021.

Ciota, A.T., 2017. West Nile virus and its vectors. Curr. Opin. Insect Sci. 22, 28–36. https://doi.org/10.1016/j.cois.2017.05.002.

Ciota, A.T., 2019. The role of co-infection and swarm dynamics in arbovirus transmission. Virus Res. 265, 88–93. https://doi.org/10.1016/j.virusres.2019.03.010.

Ciota, A.T., Keyel, A.C., 2019. The role of temperature in transmission of zoonotic arboviruses. Viruses 11, 1013. https://doi.org/10.3390/v11111013.

Ciota, A.T., Kramer, L.D., 2013. Vector-virus interactions and transmission dynamics of West Nile virus. Viruses 5, 3021–3047. https://doi.org/10.3390/v5123021.

Ciota, A.T., Drummond, C.L., Drobnack, J., Ruby, M.A., Kramer, L.D., Ebel, G.D., 2011a. Emergence of Culex pipiens from overwintering hibernacula. J Am Mosq. Contr 27, 21–29. https://doi.org/10.2987/8756-971x-27.1.21.

Ciota, A.T., Styer, L.M., Meola, M.A., Kramer, L.D., 2011b. The costs of infection and resistance as determinants of West Nile virus susceptibility in Culex mosquitoes. BMC Ecol. 11, 23. https://doi.org/10.1186/1472-6785-11-23.

Ciota, A.T., Ehrbar, D.J., Slyke, G.A.V., Payne, A.F., Willsey, G.G., Viscio, R.E., Kramer, L.D., 2012. Quantification of intrahost bottlenecks of West Nile virus in Culex pipiens mosquitoes using an artificial mutant swarm. Infect. Genet. Evol. 12, 557–564. https://doi.org/10.1016/j.meegid.2012.01.022.

Ciota, A.T., Matacchiero, A.C., Kilpatrick, A.M., Kramer, L.D., 2014. The effect of temperature on life history traits of Culex mosquitoes. J. Med. Entomol. 51, 55–62. https://doi.org/10.1603/me13003.

Ciota, A.T., Bialosuknia, S.M., Zink, S.D., Brecher, M., Ehrbar, D.J., Morrissette, M.N., Kramer, L.D., 2017. Effects of zika virus strain and Aedes mosquito species on vector competence - volume 23, number 7—July 2017 - emerging infectious diseases journal - CDC. Emerg. Infect. Dis. 23, 1110–1117. https://doi.org/10.3201/eid2307.161633.

Ciota, A.T., Chin, P.A., Ehrbar, D.J., Micieli, M.V., Fonseca, D.M., Kramer, L.D., 2018. Differential effects of temperature and mosquito genetics determine transmissibility of arboviruses by Aedes aegypti in Argentina. Am. J. Trop. Med. Hyg. 99, 417–424. https://doi.org/10.4269/ajtmh.18-0097.

Colpitts, T.M., Conway, M.J., Montgomery, R.R., Fikrig, E., 2012. West Nile virus: biology, transmission, and human infection. Clin. Microbiol. Rev. 25, 635–648. https://doi.org/10.1128/cmr.00045-12.

Conley, A.K., Fuller, D.O., Haddad, N., Hassan, A.N., Gad, A.M., Beier, J.C., 2014. Modeling the distribution of the West Nile and Rift Valley fever vector Culex pipiens in arid and semi-arid regions of the Middle East and North Africa. Parasit Vectors 7, 289. https://doi.org/10.1186/1756-3305-7-289.

Conte, A., Candeloro, L., Ippoliti, C., Monaco, F., Massis, F.D., Bruno, R., Sabatino, D.D., Danzetta, M.L., Benjelloun, A., Belkadi, B., Harrak, M.E., Declich, S., Rizzo, C., Hammami, S., Hassine, T.B., Calistri, P., Savini, G., 2015. Spatio-temporal identification of areas suitable for West Nile disease in the Mediterranean Basin and Central Europe. Plos One 10, e0146024. https://doi.org/10.1371/journal.pone.0146024.

Coon, K.L., Vogel, K.J., Brown, M.R., Strand, M.R., 2014. Mosquitoes rely on their gut microbiota for development. Mol. Ecol. 23, 2727–2739. https://doi.org/10.1111/mec.12771.

Coon, K.L., Brown, M.R., Strand, M.R., 2016. Mosquitoes host communities of bacteria that are essential for development but vary greatly between local habitats. Mol. Ecol. 25, 5806–5826. https://doi.org/10.1111/mec.13877.

Coon, K.L., Hegde, S., Hughes, G.L., 2022. Interspecies microbiome transplantation recapitulates microbial acquisition in mosquitoes. Microbiome 10, 58. https://doi.org/10.1186/s40168-022-01256-5.

Cornel, A.J., Jupp, P.G., Blackburn, N.K., 1993. Environmental temperature on the vector competence of Culex univittatus (Diptera: Culicidae) for West Nile virus. J. Med. Entomol. 30, 449–456. https://doi.org/10.1093/jmedent/30.2.449.

Cotar, A.I., Falcuta, E., Prioteasa, L.F., Dinu, S., Ceianu, C.S., Paz, S., 2016. Transmission dynamics of the West Nile virus in mosquito vector populations under the influence of weather factors in the Danube Delta, Romania. Ecohealth 13, 796–807. https://doi.org/10.1007/s10393-016-1176-y.

Crans, W.J., 2004. A classification system for mosquito life cycles: life cycle types for mosquitoes of the northeastern United States. J Vector Ecol J Soc Vector Ecol 29, 1–10.

Damos, P., Caballero, P., 2021. Detecting seasonal transient correlations between populations of the West Nile virus vector Culex sp. and temperatures with wavelet coherence analysis. Eco. Inform. 61, 101216. https://doi.org/10.1016/j.ecoinf.2021.101216.

Danforth, M.E., Reisen, W.K., Barker, C.M., 2015. Extrinsic incubation rate is not accelerated in recent California strains of West Nile virus in Culex tarsalis (Diptera: Culicidae). J. Med. Entomol. 52, 1083–1089. https://doi.org/10.1093/jme/tjv082.

Danforth, M.E., Reisen, W.K., Barker, C.M., 2016. The impact of cycling temperature on the transmission of West Nile virus. J. Med. Entomol. 53, 681–686. https://doi.org/10.1093/jme/tjw013.

Davis, C.T., Ebel, G.D., Lanciotti, R.S., Brault, A.C., Guzman, H., Siirin, M., Lambert, A., Parsons, R.E., Beasley, D.W.C., Novak, R.J., Elizondo-Quiroga, D., Green, E.N., Young, D.S., Stark, L.M., Drebot, M.A., Artsob, H., Tesh, R.B., Kramer, L.D., Barrett, A.D.T., 2005. Phylogenetic analysis of north American West Nile virus isolates, 2001–2004: evidence for the emergence of a dominant genotype. Virology 342, 252–265. https://doi.org/10.1016/j.virol.2005.07.022.

Davis, J.K., Vincent, G., Hildreth, M.B., Kightlinger, L., Carlson, C., Wimberly, M.C., 2017. Integrating environmental monitoring and mosquito surveillance to predict vector-borne disease: prospective forecasts of a West Nile virus outbreak. Plos Curr 9. https://doi.org/10.1371/currents.outbreaks.90e80717c4e67e1a830f17feeaaf85de.

Davis, J.K., Vincent, G.P., Hildreth, M.B., Kightlinger, L., Carlson, C., Wimberly, M.C., 2018. Improving the prediction of arbovirus outbreaks: a comparison of climate-driven models for West Nile virus in an endemic region of the United States. Acta Trop. 185, 242–250. https://doi.org/10.1016/j.actatropica.2018.04.028.

Dawson, J.R., Stone, W.B., Ebel, G.D., Young, D.S., Galinski, D.S., Pensabene, J.P., Franke, M.A., Eidson, M., Kramer, L.D., 2007. Crow deaths caused by West Nile virus during winter - volume 13, number 12—December 2007 - emerging infectious diseases journal - CDC. Emerg. Infect. Dis. 13, 1912–1914. https://doi.org/10.3201/eid1312.070413.

Day, J.F., Curtis, G.A., 1989. Influence of rainfall on Culex nigripalpus (Diptera: Culicidae) blood-feeding behavior in Indian River County, Florida. Ann. Entomol. Soc. Am. 82, 32–37. https://doi.org/10.1093/aesa/82.1.32.

Dellicour, S., Lequime, S., Vrancken, B., Gill, M.S., Bastide, P., Gangavarapu, K., Matteson, N.L., Tan, Y., Plessis, L.d., Fisher, A.A., Nelson, M.I., Gilbert, M., Suchard, M.A., Andersen, K.G., Grubaugh, N.D., Pybus, O.G., Lemey, P., 2020. Epidemiological hypothesis testing using a phylogeographic and phylodynamic framework. Nat. Commun. 11, 5620. https://doi.org/10.1038/s41467-020-19122-z.

Dennison, N.J., Jupatanakul, N., Dimopoulos, G., 2014. The mosquito microbiota influences vector competence for human pathogens. Curr. Opin. Insect Sci. 3, 6–13. https://doi.org/10.1016/j.cois.2014.07.004.

Didion, E.M., Doyle, M., Benoit, J.B., 2021. Bacterial communities of lab and Field northern house mosquitoes (Diptera: Culicidae) throughout diapause. J. Med. Entomol. 59, 648–658. https://doi.org/10.1093/jme/tjab184.

Dietrich, E.A., Langevin, S.A., Huang, C.Y.-H., Maharaj, P.D., Delorey, M.J., Bowen, R.A., Kinney, R.M., Brault, A.C., 2016. West Nile virus temperature sensitivity and avian virulence are modulated by NS1-2B polymorphisms. Plos Neglect. Trop. D 10, e0004938. https://doi.org/10.1371/journal.pntd.0004938.

Dietz, K., 1993. The estimation of the basic reproduction number for infectious diseases. Stat. Methods Med. Res. 2, 23–41. https://doi.org/10.1177/096228029300200103.

Dodson, B.L., Kramer, L.D., Rasgon, J.L., 2011. Larval nutritional stress does not affect vector competence for West Nile virus (WNV) in Culex tarsalis. Vector-borne Zoonot 11, 1493–1497. https://doi.org/10.1089/vbz.2011.0662.

Dodson, B.L., Kramer, L.D., Rasgon, J.L., 2012. Effects of larval rearing temperature on immature development and West Nile virus vector competence of Culex tarsalis. Parasit Vectors 5, 199. https://doi.org/10.1186/1756-3305-5-199.

Dodson, B.L., Hughes, G.L., Paul, O., Matacchiero, A.C., Kramer, L.D., Rasgon, J.L., 2014. Wolbachia enhances West Nile virus (WNV) infection in the mosquito Culex tarsalis. Plos Neglect. Trop. D 8, e2965. https://doi.org/10.1371/journal.pntd.0002965.

Dohm, D.J., O'Guinn, M.L., Turell, M.J., 2002. Effect of environmental temperature on the ability of Culex pipiens (Diptera: Culicidae) to transmit West Nile virus. J. Med. Entomol. 39, 221–225. https://doi.org/10.1603/0022-2585-39.1.221.

Duguma, D., Hall, M.W., Smartt, C.T., Neufeld, J.D., 2017. Effects of organic amendments on microbiota associated with the Culex nigripalpus mosquito vector of the Saint Louis encephalitis and West Nile viruses. Msphere 2, e00387–16. https://doi.org/10.1128/msphere.00387-16.

Dutra, H.L.C., Rocha, M.N., Dias, F.B.S., Mansur, S.B., Caragata, E.P., Moreira, L.A., 2016. Wolbachia blocks currently circulating zika virus isolates in Brazilian Aedes aegypti mosquitoes. Cell Host Microbe 19, 771–774. https://doi.org/10.1016/j.chom.2016.04.021.

Ebel, G.D., Carricaburu, J., Young, D., Bernard, K., Kramer, L., 2004. Genetic and phenotypic variation of West Nile virus in New York, 2000-2003. Am. J. Trop. Med. Hyg. 71, 493–500. https://doi.org/10.4269/ajtmh.2004.71.493.

ECDC, 2022. West Nile Virus Infection [WWW Document]. URL https://www.ecdc.europa.eu/en/west-nile-virus-infection. (accessed 22).

Ehrbar, D.J., Ngo, K.A., Campbell, S.R., Kramer, L.D., Ciota, A.T., 2017. High levels of local inter- and intra-host genetic variation of West Nile virus and evidence of fine-scale evolutionary pressures. Infect. Genet. Evol. 51, 219–226. https://doi.org/10.1016/j.meegid.2017.04.010.

Ekwudu, O., Marquart, L., Webb, L., Lowry, K.S., Devine, G.J., Hugo, L.E., Frentiu, F.D., 2020. Effect of serotype and strain diversity on dengue virus replication in Australian mosquito vectors. Pathogens 9, 668. https://doi.org/10.3390/pathogens9080668.

Eldridge, B.F., 1968. The effect of temperature and photoperiod on blood-feeding and ovarian development in mosquitoes of the Culex Pipiens complex. Am. J. Trop. Med. Hyg. 17, 133–140. https://doi.org/10.4269/ajtmh.1968.17.133.

Epstein, P.R., Defilippo, C., 2001. West Nile virus and drought. Global Change Hum Heal 2, 105–107. https://doi.org/10.1023/a:1015089901425.

Farajollahi, A., Crans, W.J., Nickerson, D., Bryant, P., Wolf, B., Glaser, A., Andreadis, T.G., 2005. Detection of West Nile virus RNA from the louse fly Icosta americana (Diptera: Hippoboscidae). J Am Mosq. Contr 21, 474–476. https://doi.org/10.2987/8756-971x(2006)21[474:downvr]2.0.co;2.

Farajollahi, A., Fonseca, D.M., Kramer, L.D., Kilpatrick, A.M., 2011. "Bird biting" mosquitoes and human disease: a review of the role of Culex pipiens complex mosquitoes in epidemiology. Infect. Genet. Evol. 11, 1577–1585. https://doi.org/10.1016/j.meegid.2011.08.013.

Fay, R.L., Ngo, K.A., Kuo, L., Willsey, G.G., Kramer, L.D., Ciota, A.T., 2021. Experimental evolution of West Nile virus at higher temperatures facilitates broad adaptation and increased genetic diversity. Viruses 13, 1889. https://doi.org/10.3390/v13101889.

Ferreira, P.G., Tesla, B., Horácio, E.C.A., Nahum, L.A., Brindley, M.A., Mendes, T.A.d. O., Murdock, C.C., 2020. Temperature dramatically shapes mosquito gene expression with consequences for mosquito–zika virus interactions. Front. Microbiol. 11, 901. https://doi.org/10.3389/fmicb.2020.00901.

Franklinos, L.H.V., Jones, K.E., Redding, D.W., Abubakar, I., 2019. The effect of global change on mosquito-borne disease. Lancet Infect. Dis. 19, e302–e312. https://doi.org/10.1016/s1473-3099(19)30161-6.

Gale, P., 2020. How virus size and attachment parameters affect the temperature sensitivity of virus binding to host cells: predictions of a thermodynamic model for arboviruses and HIV. Microb. Risk Anal. 15, 100104. https://doi.org/10.1016/j.mran.2020.100104.

Gardner, A.M., Hamer, G.L., Hines, A.M., Newman, C.M., Walker, E.D., Ruiz, M.O., 2012. Weather variability affects abundance of larval Culex (Diptera: Culicidae) in storm water catch basins in suburban Chicago. J. Med. Entomol. 49, 270–276. https://doi.org/10.1603/me11073.

Glaser, R.L., Meola, M.A., 2010. The native Wolbachia endosymbionts of Drosophila melanogaster and Culex quinquefasciatus increase host resistance to West Nile virus infection. Plos One 5, e11977. https://doi.org/10.1371/journal.pone.0011977.

Gloria-Soria, A., Armstrong, P.M., Powell, J.R., Turner, P.E., 2017. Infection rate of Aedes aegypti mosquitoes with dengue virus depends on the interaction between temperature and mosquito genotype. Proc Royal Soc B Biol. Sci 284, 20171506. https://doi.org/10.1098/rspb.2017.1506.

Gloria-Soria, A., Payne, A.F., Bialosuknia, S.M., Stout, J., Mathias, N., Eastwood, G., Ciota, A.T., Kramer, L.D., Armstrong, P.M., 2020. Vector competence of Aedes albopictus populations from the northeastern United States for chikungunya, dengue, and zika viruses. Am. J. Trop. Med. Hyg. https://doi.org/10.4269/ajtmh.20-0874.

Goldberg, T.L., Anderson, T.K., Hamer, G.L., 2010. West Nile virus may have hitched a ride across the Western United States on Culex tarsalis mosquitoes. Mol. Ecol. 19, 1518–1519. https://doi.org/10.1111/j.1365-294x.2010.04578.x.

Goldblum, N., Sterk, V., Paderski, B., 1954. West Nile fever. The clinical features of the disease and the isolation of West Nile virus from the blood of nine human cases. Am. J. Hyg. 59, 89–103.

Gorris, M.E., Bartlow, A.W., Temple, S.D., Romero-Alvarez, D., Shutt, D.P., Fair, J.M., Kaufeld, K.A., Valle, S.Y.D., Manore, C.A., 2021. Updated distribution maps of predominant Culex mosquitoes across the Americas. Parasit Vectors 14, 547. https://doi.org/10.1186/s13071-021-05051-3.

Grech, M.G., Sartor, P.D., Almirón, W.R., Ludueña-Almeida, F.F., 2015. Effect of temperature on life history traits during immature development of Aedes aegypti and Culex quinquefasciatus (Diptera: Culicidae) from Córdoba city, Argentina. Acta Trop. 146, 1–6. https://doi.org/10.1016/j.actatropica.2015.02.010.

Grimstad, P.R., Haramis, L.D., 1984. Aedes Triseriatus (Diptera: Culicidae) and La Crosse virus III. Enhanced Oral transmission by nutrition-deprived Mosquitoes1. J. Med. Entomol. 21, 249–256. https://doi.org/10.1093/jmedent/21.3.249.

Gunay, F., Alten, B., Ozsoy, E.D., 2010. Estimating reaction norms for predictive population parameters, age specific mortality, and mean longevity in temperature-dependent cohorts of Culex quinquefasciatus say (Diptera: Culicidae). J. Vector Ecol. 35, 354–362. https://doi.org/10.1111/j.1948-7134.2010.00094.x.

Hahn, M.B., Monaghan, A.J., Hayden, M.H., Eisen, R.J., Delorey, M.J., Lindsey, N.P., Nasci, R.S., Fischer, M., 2015. Meteorological conditions associated with increased incidence of West Nile virus disease in the United States, 2004–2012. Am. J. Trop. Med. Hyg. 92, 1013–1022. https://doi.org/10.4269/ajtmh.14-0737.

Hamer, G.L., Kitron, U.D., Goldberg, T.L., Brawn, J.D., Loss, S.R., Ruiz, M.O., Hayes, D.B., Walker, E.D., 2009. Host selection by Culex pipiens mosquitoes and

West Nile virus amplification. Am. J. Trop. Med. Hyg. 80, 268–278. https://doi.org/10.4269/ajtmh.2009.80.268.

Hardy, J.L., Houk, E.J., Kramer, L.D., Reeves, W.C., 1983. Intrinsic factors affecting vector competence of mosquitoes for arboviruses. Annu. Rev. Entomol. 28, 229–262. https://doi.org/10.1146/annurev.en.28.010183.001305.

Hartemink, N.A., Davis, S.A., Reiter, P., Hublek, Z., Heesterbeek, J.A.P., 2007. Importance of bird-to-bird transmission for the establishment of West Nile virus. Vector-borne Zoonot 7, 575–584. https://doi.org/10.1089/vbz.2006.0613.

Hayes, E.B., Komar, N., Nasci, R.S., Montgomery, S.P., O'Leary, D.R., Campbell, G.L., 2005. Epidemiology and transmission dynamics of West Nile virus disease. Emerg. Infect. Dis. 11, 1167–1173. https://doi.org/10.3201/eid1108.050289a.

Hegde, S., Rasgon, J.L., Hughes, G.L., 2015. The microbiome modulates arbovirus transmission in mosquitoes. Curr. Opin. Virol. 15, 97–102. https://doi.org/10.1016/j.coviro.2015.08.011.

Hernández-Triana, L.M., Jeffries, C.L., Mansfield, K.L., Carnell, G., Fooks, A.R., Johnson, N., 2014. Emergence of West Nile virus lineage 2 in Europe: a review on the introduction and spread of a mosquito-borne disease. Front. Public Health 2, 271. https://doi.org/10.3389/fpubh.2014.00271.

Holmes, C.J., Brown, E., Sharma, D., Nguyen, Q., Spangler, A., Pathak, A., Payton, B., Warden, M., Shah, A.J., Shaw, S., Benoit, J.B., 2022. Bloodmeal regulation in mosquitoes curtails dehydration-induced mortality, altering vectorial capacity. J. Insect Physiol. 104363. https://doi.org/10.1016/j.jinsphys.2022.104363.

Hongoh, V., Berrang-Ford, L., Scott, M.E., Lindsay, L.R., 2012. Expanding geographical distribution of the mosquito, Culex pipiens, in Canada under climate change. Appl Geogr 33, 53–62. https://doi.org/10.1016/j.apgeog.2011.05.015.

Hoover, K.C., Barker, C.M., 2016. West Nile virus, climate change, and circumpolar vulnerability. Wiley Interdiscip. Rev. Clim. Change 7, 283–300. https://doi.org/10.1002/wcc.382.

Hubálek, Z., 2000. European experience with the West Nile virus ecology and epidemiology: could it be relevant for the New World? Viral Immunol. 13, 415–426. https://doi.org/10.1089/vim.2000.13.415.

Huntley, B., Collingham, Y.C., Green, R.E., Hilton, G.M., Rahbek, C.G., Willis, S.G., 2006. Potential impacts of climatic change upon geographical distributions of birds. Ibis 148, 8–28. https://doi.org/10.1111/j.1474-919x.2006.00523.x.

Hussain, M., Lu, G., Torres, S., Edmonds, J.H., Kay, B.H., Khromykh, A.A., Asgari, S., 2013. Effect of Wolbachia on replication of West Nile virus in a mosquito cell line and adult mosquitoes. J. Virol. 87, 851–858. https://doi.org/10.1128/jvi.01837-12.

Johnson, K.N., 2015. The impact of Wolbachia on virus infection in mosquitoes. Viruses 7, 5705–5717. https://doi.org/10.3390/v7112903.

Johnson, B.J., Sukhdeo, M.V.K., 2013. Drought-induced amplification of local and regional West Nile virus infection rates in New Jersey. J. Med. Entomol. 50, 195–204. https://doi.org/10.1603/me12035.

Johnson, L.R., Ben-Horin, T., Lafferty, K.D., McNally, A., Mordecai, E., Paaijmans, K.P., Pawar, S., Ryan, S.J., 2015. Understanding uncertainty in temperature effects on vector-borne disease: a Bayesian approach. Ecology 96, 203–213. https://doi.org/10.1890/13-1964.1.

Jones, C.E., Lounibos, L.P., Marra, P.P., Kilpatrick, A.M., 2012. Rainfall influences survival of Culex pipiens (Diptera: Culicidae) in a residential neighborhood in the mid-Atlantic United States. J. Med. Entomol. 49, 467–473. https://doi.org/10.1603/me11191.

Jourdain, E., Olsen, B., Lundkvist, A., Hubálek, Z., Šikutová, S., Waldenström, J., Karlsson, M., Wahlström, M., Jozan, M., Falk, K.I., 2011. Surveillance for West Nile virus in wild birds from northern Europe. Vector-borne Zoonot 11, 77–79. https://doi.org/10.1089/vbz.2009.0028.

Kain, M.P., Bolker, B.M., 2019. Predicting West Nile virus transmission in north American bird communities using phylogenetic mixed effects models and eBird citizen science data. Parasit Vectors 12, 395. https://doi.org/10.1186/s13071-019-3656-8.

Kampen, H., Tews, B.A., Werner, D., 2021. First evidence of West Nile virus overwintering in mosquitoes in Germany. Viruses 13, 2463. https://doi.org/10.3390/v13122463.

Kang, D.S., Barron, M.S., Lovin, D.D., Cunningham, J.M., Eng, M.W., Chadee, D.D., Li, J., Severson, D.W., 2018. A transcriptomic survey of the impact of environmental stress on response to dengue virus in the mosquito, Aedes aegypti. Plos Neglect. Trop. D 12, e0006568. https://doi.org/10.1371/journal.pntd.0006568.

Karki, S., Brown, W.M., Uelmen, J., Ruiz, M.O., Smith, R.L., 2020. The drivers of West Nile virus human illness in the Chicago, Illinois, USA area: fine scale dynamic effects of weather, mosquito infection, social, and biological conditions. Plos One 15, e0227160. https://doi.org/10.1371/journal.pone.0227160.

Kay, B.H., Jennings, C.D., 2002. Enhancement or modulation of the vector competence of Ochlerotatus vigilax (Diptera: Culicidae) for Ross River virus by temperature. J. Med. Entomol. 39, 99–105. https://doi.org/10.1603/0022-2585-39.1.99.

Kernbach, M.E., Martin, L.B., Unnasch, T.R., Hall, R.J., Jiang, R.H.Y., Francis, C.D., 2021. Light pollution affects West Nile virus exposure risk across Florida. Proc. R. Soc. B 288, 20210253. https://doi.org/10.1098/rspb.2021.0253.

Keyel, A.C., Timm, O.E., Backenson, P.B., Prussing, C., Quinones, S., McDonough, K.A., Vuille, M., Conn, J.E., Armstrong, P.M., Andreadis, T.G., Kramer, L.D., 2019. Seasonal temperatures and hydrological conditions improve the prediction of West Nile virus infection rates in Culex mosquitoes and human case counts in New York and Connecticut. Plos One 14, e0217854. https://doi.org/10.1371/journal.pone.0217854.

Keyel, A.C., Raghavendra, A., Ciota, A.T., Timm, O.E., 2021. West Nile virus is predicted to be more geographically widespread in New York state and Connecticut under future climate change. Glob. Chang. Biol. https://doi.org/10.1111/gcb.15842.

Kilpatrick, A.M., Wheeler, S.S., 2019. Impact of West Nile virus on bird populations: limited lasting effects, evidence for recovery, and gaps in our understanding of impacts on ecosystems. J. Med. Entomol. 56, 1491–1497. https://doi.org/10.1093/jme/tjz149.

Kilpatrick, A.M., Kramer, L.D., Campbell, S.R., Alleyne, E.O., Dobson, A.P., Daszak, P., 2005. West Nile virus risk assessment and the bridge vector paradigm - volume 11, number 3—march 2005 - emerging infectious diseases journal - CDC. Emerg. Infect. Dis. 11, 425–429. https://doi.org/10.3201/eid1103.040364.

Kilpatrick, A.M., Daszak, P., Jones, M.J., Marra, P.P., Kramer, L.D., 2006. Host heterogeneity dominates West Nile virus transmission. Proc Royal Soc B Biol. Sci 273, 2327–2333. https://doi.org/10.1098/rspb.2006.3575.

Kilpatrick, A.M., Meola, M.A., Moudy, R.M., Kramer, L.D., 2008. Temperature, viral genetics, and the transmission of West Nile virus by Culex pipiens mosquitoes. PLoS Pathog. 4, e1000092. https://doi.org/10.1371/journal.ppat.1000092.

Kinney, R.M., Huang, C.Y.-H., Whiteman, M.C., Bowen, R.A., Langevin, S.A., Miller, B.R., Brault, A.C., 2006. Avian virulence and thermostable replication of the north American strain of West Nile virus. J. Gen. Virol. 87, 3611–3622. https://doi.org/10.1099/vir.0.82299-0.

Klenk, K., Snow, J., Morgan, K., Bowen, R.A., Stephens, M., Foster, F., Gordy, P., Beckett, S., Komar, N., Gubler, D., Bunning, M.L., 2004. Alligators as West Nile virus amplifiers - volume 10, number 12—December 2004 - emerging infectious diseases journal - CDC. Emerg. Infect. Dis. 10, 2150–2155. https://doi.org/10.3201/eid1012.040264.

Koenraadt, C.J.M., Harrington, L.C., 2008. Flushing effect of rain on container-inhabiting mosquitoes Aedes aegypti and Culex pipiens (Diptera: Culicidae). J. Med. Entomol. 45, 28–35. https://doi.org/10.1093/jmedent/45.1.28.

Komar, N., Langevin, S., Hinten, S., Nemeth, N.M., Edwards, E., Hettler, D.L., Davis, B.S., Bowen, R.A., Bunning, M.L., 2003. Experimental infection of north American birds with the New York 1999 strain of West Nile virus - volume 9, number 3—march 2003 - emerging infectious diseases journal - CDC. Emerg. Infect. Dis. 9, 311–322. https://doi.org/10.3201/eid0903.020628.

Komar, N., Panella, N.A., Young, G.R., Brault, A.C., Levy, C.E., 2013. Avian hosts of West Nile virus in Arizona. Am. J. Trop. Med. Hyg. 89, 474–481. https://doi.org/10.4269/ajtmh.13-0061.

Kramer, L.D., Hardy, J.L., Presser, S.B., 1998. Characterization of modulation of Western equine encephalomyelitis virus by Culex tarsalis (Diptera: Culicidae) maintained at 32°C following parenteral infection. J. Med. Entomol. 35, 289–295. https://doi.org/10.1093/jmedent/35.3.289.

Kramer, L.D., Ciota, A.T., Kilpatrick, A.M., 2019. Introduction, spread, and establishment of West Nile virus in the Americas. J. Med. Entomol. 56, 1448–1455. https://doi.org/10.1093/jme/tjz151.

Lambrechts, L., Paaijmans, K.P., Fansiri, T., Carrington, L.B., Kramer, L.D., Thomas, M.B., Scott, T.W., 2011. Impact of daily temperature fluctuations on dengue virus transmission by Aedes aegypti. Proc National Acad Sci 108, 7460–7465. https://doi.org/10.1073/pnas.1101377108.

Lanciotti, R.S., Roehrig, J.T., Deubel, V., Smith, J., Parker, M., Steele, K., Crise, B., Volpe, K.E., Crabtree, M.B., Scherret, J.H., Hall, R.A., MacKenzie, J.S., Cropp, C.B., Panigrahy, B., Ostlund, E., Schmitt, B., Malkinson, M., Banet, C., Weissman, J., Komar, N., Savage, H.M., Stone, W., McNamara, T., Gubler, D.J., 1999. Origin of the West Nile virus responsible for an outbreak of encephalitis in the northeastern United States. Science 286, 2333–2337. https://doi.org/10.1126/science.286.5448.2333.

Landesman, W.J., Allan, B.F., Langerhans, R.B., Knight, T.M., Chase, J.M., 2007. Inter-annual associations between precipitation and human incidence of West Nile virus in the United States. Vector Borne Zoonotic Dis. 7 (3), 337–343. https://doi.org/10.1089/vbz.2006.0590.

Lauring, A.S., Frydman, J., Andino, R., 2013. The role of mutational robustness in RNA virus evolution. Nat. Rev. Microbiol. 11, 327–336. https://doi.org/10.1038/nrmicro3003.

Lebl, K., Brugger, K., Rubel, F., 2013. Predicting Culex pipiens/restuans population dynamics by interval lagged weather data. Parasit Vectors 6, 129. https://doi.org/10.1186/1756-3305-6-129.

Lee, W.-S., Webster, J.A., Madzokere, E.T., Stephenson, E.B., Herrero, L.J., 2019. Mosquito antiviral defense mechanisms: a delicate balance between innate immunity and persistent viral infection. Parasit Vectors 12, 165. https://doi.org/10.1186/s13071-019-3433-8.

Levine, R.S., Mead, D.G., Hamer, G.L., Brosi, B.J., Hedeen, D.L., Hedeen, M.W., McMillan, J.R., Bisanzio, D., Kitron, U.D., 2016. Supersuppression: reservoir competency and timing of mosquito host shifts combine to reduce spillover of West Nile virus. Am. J. Trop. Med. Hyg. 95, 1174–1184. https://doi.org/10.4269/ajtmh.15-0809.

Levine, R.S., Hedeen, D.L., Hedeen, M.W., Hamer, G.L., Mead, D.G., Kitron, U.D., 2017. Avian species diversity and transmission of West Nile virus in Atlanta, Georgia. Parasit Vectors 10, 62. https://doi.org/10.1186/s13071-017-1999-6.

Lipsitch, M., Moxon, E.R., 1997. Virulence and transmissibility of pathogens: what is the relationship? Trends Microbiol. 5, 31–37. https://doi.org/10.1016/s0966-842x(97)81772-6.

Little, E., Campbell, S.R., Shaman, J., 2016. Development and validation of a climate-based ensemble prediction model for West Nile virus infection rates in Culex mosquitoes,

Suffolk County, New York. Parasit Vectors 9, 443. https://doi.org/10.1186/s13071-016-1720-1.
Liu, Z., Xu, Y., Li, Y., Xu, S., Li, Y., Xiao, L., Chen, X., He, C., Zheng, K., 2022. Transcriptome analysis of Aedes albopictus midguts infected by dengue virus identifies a gene network module highly associated with temperature. Parasit Vectors 15, 173. https://doi.org/10.1186/s13071-022-05282-y.
Loetti, V., Schweigmann, N., Burroni, N., 2011a. Development rates, larval survivorship and wing length of Culex pipiens (Diptera: Culicidae) at constant temperatures. J. Nat. Hist. 45, 2203–2213. https://doi.org/10.1080/00222933.2011.590946.
Loetti, V., Schweigmann, N., Burroni, N., 2011b. Temperature effects on the immature development time of Culex eduardoi Casal & García (Diptera: Culicidae). Neotrop. Entomol. 40, 138–142. https://doi.org/10.1590/s1519-566x2011000100021.
Lourenço, J., Thompson, R.N., Thézé, J., Obolski, U., 2020. Characterising West Nile virus epidemiology in Israel using a transmission suitability index. Eurosurveillance 25, 1900629. https://doi.org/10.2807/1560-7917.es.2020.25.46.1900629.
Lourenço, J., Barros, S.C., Zé-Zé, L., Damineli, D.S.C., Giovanetti, M., Osório, H.C., Amaro, F., Henriques, A.M., Ramos, F., Luís, T., Duarte, M.D., Fagulha, T., Alves, M.J., Obolski, U., 2022. West Nile virus transmission potential in Portugal. Commun Biol. 5, 6. https://doi.org/10.1038/s42003-021-02969-3.
Macdonald, G., 1961. Epidemiologic models in studies of vector-borne diseases: the R. E. Dyer lecture. Public Health Rep. 1896–1970 (76), 753. https://doi.org/10.2307/4591271.
Manore, C.A., Davis, J.K., Christofferson, R.C., Wesson, D.M., Hyman, J.M., Mores, C.N., 2014. Towards an early warning system for forecasting human West Nile virus incidence. Plos Curr 6. https://doi.org/10.1371/currents.outbreaks.f0b3978230599a56830ce30cb9ce0500.
Marcantonio, M., Rizzoli, A., Metz, M., Rosà, R., Marini, G., Chadwick, E., Neteler, M., 2015. Identifying the environmental conditions Favouring West Nile virus outbreaks in Europe. Plos One 10, e0121158. https://doi.org/10.1371/journal.pone.0121158.
Marini, G., Poletti, P., Giacobini, M., Pugliese, A., Merler, S., Rosà, R., 2016. The role of climatic and density dependent factors in shaping mosquito population dynamics: the case of Culex pipiens in northwestern Italy. Plos One 11, e0154018. https://doi.org/10.1371/journal.pone.0154018.
Marini, G., Calzolari, M., Angelini, P., Bellini, R., Bellini, S., Bolzoni, L., Torri, D., Defilippo, F., Dorigatti, I., Nikolay, B., Pugliese, A., Rosà, R., Tamba, M., 2020a. A quantitative comparison of West Nile virus incidence from 2013 to 2018 in Emilia-Romagna, Italy. Plos Neglect. Trop. D 14, e0007953. https://doi.org/10.1371/journal.pntd.0007953.
Marini, G., Manica, M., Delucchi, L., Pugliese, A., Rosà, R., 2020b. Spring temperature shapes West Nile virus transmission in Europe. Acta Trop. 215, 105796. https://doi.org/10.1016/j.actatropica.2020.105796.
Martinez, J., Longdon, B., Bauer, S., Chan, Y.-S., Miller, W.J., Bourtzis, K., Teixeira, L., Jiggins, F.M., 2014. Symbionts commonly provide broad Spectrum resistance to viruses in insects: a comparative analysis of Wolbachia strains. PLoS Pathog. 10, e1004369. https://doi.org/10.1371/journal.ppat.1004369.
Marzluff, J.M., 2001. Avian Ecology and Conservation in an Urbanizing World 19–47. https://doi.org/10.1007/978-1-4615-1531-9_2.
McMullen, A.R., May, F.J., Li, L., Guzman, H., Bueno, R., Dennett, J.A., Tesh, R.B., Barrett, A.D.T., 2011. Evolution of new genotype of West Nile virus in North America. Emerg. Infect. Dis. 17 (5), 785–793. https://doi.org/10.3201/eid1705.101707.
Melnick, J.L., Paul, J.R., Riordan, J.T., Barnett, V.H., Goldblum, N., Zabin, E., 1951. Isolation from human sera in Egypt of a virus apparently identical to West Nile

virus.*. Proc. Soc. Exp. Biol. Med. 77, 661–665. https://doi.org/10.3181/00379727-77-18884.
Mercier, A., Obadia, T., Carraretto, D., Velo, E., Gabiane, G., Bino, S., Vazeille, M., Gasperi, G., Dauga, C., Malacrida, A.R., Reiter, P., Failloux, A.-B., 2022. Impact of temperature on dengue and chikungunya transmission by the mosquito Aedes albopictus. Sci. Rep. 12, 6973. https://doi.org/10.1038/s41598-022-10977-4.
Micieli, M.V., Glaser, R.L., 2014. Somatic Wolbachia (Rickettsiales: Rickettsiaceae) levels in Culex quinquefasciatus and Culex pipiens (Diptera: Culicidae) and resistance to West Nile virus infection. J. Med. Entomol. 51, 189–199. https://doi.org/10.1603/me13152.
Mihailović, D.T., Petrić, D., Petrović, T., Hrnjaković-Cvjetković, I., Djurdjevic, V., Nikolić-Đorić, E., Arsenić, I., Petrić, M., Mimić, G., Ignjatović-Ćupina, A., 2020. Assessment of climate change impact on the malaria vector Anopheles hyrcanus, West Nile disease, and incidence of melanoma in the Vojvodina Province (Serbia) using data from a regional climate model. Plos One 15, e0227679. https://doi.org/10.1371/journal.pone.0227679.
Moise, I.K., Riegel, C., Muturi, E.J., 2018. Environmental and social-demographic predictors of the southern house mosquito Culex quinquefasciatus in New Orleans, Louisiana. Parasit Vectors 11, 249. https://doi.org/10.1186/s13071-018-2833-5.
Montecino-Latorre, D., Barker, C.M., 2018. Overwintering of West Nile virus in a bird community with a communal crow roost. Sci. Rep. 8, 6088. https://doi.org/10.1038/s41598-018-24133-4.
Mordecai, E.A., Caldwell, J.M., Grossman, M.K., Lippi, C.A., Johnson, L.R., Neira, M., Rohr, J.R., Ryan, S.J., Savage, V., Shocket, M.S., Sippy, R., Ibarra, A.M.S., Thomas, M.B., Villena, O., 2019. Thermal biology of mosquito-borne disease. Ecol. Lett. 22, 1690–1708. https://doi.org/10.1111/ele.13335.
Morin, C.W., Comrie, A.C., 2010. Modeled response of the West Nile virus vector Culex quinquefasciatus to changing climate using the dynamic mosquito simulation model. Int. J. Biometeorol. 54, 517–529. https://doi.org/10.1007/s00484-010-0349-6.
Moudy, R.M., Meola, M.A., Morin, L.-L.L., Ebel, G.D., Kramer, L.D., 2007. A newly emergent genotype of West Nile virus is transmitted earlier and more efficiently by Culex mosquitoes. Am. J. Trop. Med. Hyg. 77, 365–370.
Mpho, M., Callaghan, A., Holloway, G.J., 2002. Temperature and genotypic effects on life history and fluctuating asymmetry in a field strain of Culex pipiens. Heredity 88, 307–312. https://doi.org/10.1038/sj.hdy.6800045.
Murrieta, R.A., Garcia-Luna, S.M., Murrieta, D.J., Halladay, G., Young, M.C., Fauver, J.R., Gendernalik, A., Weger-Lucarelli, J., Rückert, C., Ebel, G.D., 2021. Impact of extrinsic incubation temperature on natural selection during zika virus infection of Aedes aegypti and Aedes albopictus. PLoS Pathog. 17, e1009433. https://doi.org/10.1371/journal.ppat.1009433.
Muturi, E.J., Nyakeriga, A., Blackshear, M., 2012. Temperature-mediated differential expression of immune and stress-related genes in Aedes aegypti larvae. J Am Mosq. Contr 28, 79–83. https://doi.org/10.2987/11-6194r.1.
Myer, M.H., Johnston, J.M., 2019. Spatiotemporal Bayesian modeling of West Nile virus: identifying risk of infection in mosquitoes with local-scale predictors. Sci. Total Environ. 650, 2818–2829. https://doi.org/10.1016/j.scitotenv.2018.09.397.
Myer, M.H., Campbell, S.R., Johnston, J.M., 2017. Spatiotemporal modeling of ecological and sociological predictors of West Nile virus in Suffolk County, NY, mosquitoes. Ecosphere Wash D C 8, e01854. https://doi.org/10.1002/ecs2.1854.
Nasci, R.S., Savage, H.M., White, D.J., Miller, J.R., Cropp, B.C., Godsey, M.S., Kerst, A.J., Bennett, P., Gottfried, K., Lanciotti, R.S., 2001. West Nile virus in overwintering Culex mosquitoes, new York City, 2000. Emerg. Infect. Dis. 7, 742–744. https://doi.org/10.3201/eid0704.010426.

Ngo, K.A., Rose, J.T., Kramer, L.D., Ciota, A.T., 2019. Adaptation of Rabensburg virus (RBGV) to vertebrate hosts by experimental evolution. Virology 528, 30–36. https://doi.org/10.1016/j.virol.2018.11.015.

Novakova, E., Woodhams, D.C., Rodríguez-Ruano, S.M., Brucker, R.M., Leff, J.W., Maharaj, A., Amir, A., Knight, R., Scott, J., 2017. Mosquito microbiome dynamics, a background for prevalence and seasonality of West Nile virus. Front. Microbiol. 8, 526. https://doi.org/10.3389/fmicb.2017.00526.

Oda, T., Mori, A., Kurokawa, K., 1980. Effects of temperatures on the oviposition and hatching of eggs in Culex pipiens molestus and Culex pipiens quinquefasciatus. Trop. Med 22, 167–180.

Oda, T., Uchida, K., Mori, A., Mine, M., Eshita, Y., Kurokawa, K., Kato, K., Tahara, H., 1999. Effects of high temperature on the emergence and survival of adult Culex pipiens molestus and Culex quinquefasciatus in Japan. J Am Mosq. Contr 15, 153–156.

Olive, M.-M., Broban, A., Andriamandimby, S.F., Dorsemans, A.-C., Ravalohery, J.-P., Andriamamonjy, S., Rakotomanana, F., Rogier, C., Heraud, J.-M., 2021. Seroprevalence and Risk Factors Associated with West Nile Infection in Human in Madagascar: A Cross-Sectional Serological Survey. https://doi.org/10.21203/rs.3.rs-388847/v1.

Onyango, M., Payne, A., Stout, J., Dieme, C., Kuo, L., Ciota, A., Kramer, L., 2020a. Potential for transmission of Elizabethkingia anophelis by Aedes albopictus and the role of microbial interactions in zika virus competence. Biorxiv 702464. https://doi.org/10.1101/702464.

Onyango, M.G., Bialosuknia, S.M., Payne, A.F., Mathias, N., Kuo, L., Vigneron, A., DeGennaro, M., Ciota, A.T., Kramer, L.D., 2020b. Increased temperatures reduce the vectorial capacity of Aedes mosquitoes for zika virus. Emerg Microbes Infec 9, 67–77. https://doi.org/10.1080/22221751.2019.1707125.

Owen, J., Moore, F., Panella, N., Edwards, E., Bru, R., Hughes, M., Komar, N., 2006. Migrating birds as dispersal vehicles for West Nile virus. Ecohealth 3, 79. https://doi.org/10.1007/s10393-006-0025-9.

Owen, J.C., Landwerlen, H.R., Dupuis, A.P., Belsare, A.V., Sharma, D.B., Wang, S., Ciota, A.T., Kramer, L.D., 2021. Reservoir hosts experiencing food stress alter transmission dynamics for a zoonotic pathogen. Proc Royal Soc B Biol. Sci 288, 20210881. https://doi.org/10.1098/rspb.2021.0881.

Pachauri, R.K., Allen, M.R., Barros, V.R., Broome, J., Cramer, W., Christ, R., Church, J.A., Clarke, L., Dahe, Q., Dasgupta, P., Dubash, N.K., Edenhofer, O., Elgizouli, I., Field, C.B., Forster, P., Friedlingstein, P., Fuglestvedt, J., Gomez-Echeverri, L., Hallegatte, S., Hegerl, G., Howden, M., Jiang, K., Cisneros, B.J., Kattsov, V., Lee, H., Mach, K.J., Marotzke, J., Mastrandrea, M.D., Meyer, L., Minx, J., Mulugetta, Y., O'Brien, K., Oppenheimer, M., Pereira, J.J., Pichs-Madruga, R., Plattner, G.-K., Pörtner, H.-O., Power, S.B., Preston, B., Ravindranath, N.H., Reisinger, A., Riahi, K., Rusticucci, M., Scholes, R., Seyboth, K., Sokona, Y., Stavins, R., Stocker, T.F., Tschakert, P., Vuuren, D.v., Ypersele, J.-P.v., 2014. Fifth Assessment Report of the Intergovernmental Panel on Climate Change. In IPCC Climate Change 2014: Synthesis Report [WWW Document]. URL https://epic.awi.de/id/eprint/37530/1/IPCC_AR5_SYR_Final.pdf. (accessed 9.20.21).

Paradkar, P.N., Trinidad, L., Voysey, R., Duchemin, J.-B., Walker, P.J., 2012. Secreted Vago restricts West Nile virus infection in Culex mosquito cells by activating the jak-STAT pathway. Proc National Acad Sci 109, 18915–18920. https://doi.org/10.1073/pnas.1205231109.

Paradkar, P.N., Duchemin, J.-B., Voysey, R., Walker, P.J., 2014. Dicer-2-dependent activation of Culex Vago occurs via the TRAF-Rel2 signaling pathway. Plos Neglect. Trop. D 8, e2823. https://doi.org/10.1371/journal.pntd.0002823.

Patrick, M.L., Bradley, T.J., 2000. The physiology of salinity tolerance in larvae of two species of Culex mosquitoes: the role of compatible solutes. J. Exp. Biol. 203, 821–830. https://doi.org/10.1242/jeb.203.4.821.

Paulson, S.L., Hawley, W.A., 1991. Effect of body size on the vector competence of field and laboratory populations of Aedes triseriatus for La Crosse virus. J. Am. Mosq. Control Assoc. 7, 170–175.

Paz, S., Semenza, J.C., 2013. Environmental drivers of West Nile fever epidemiology in Europe and Western Asia—a review. Int. J. Environ. Res. Pu 10, 3543–3562. https://doi.org/10.3390/ijerph10083543.

Pérez-Ramírez, E., Llorente, F., Jiménez-Clavero, M.Á., 2014. Experimental infections of wild birds with West Nile virus. Viruses 6, 752–781. https://doi.org/10.3390/v6020752.

Platonov, A.E., Fedorova, M.V., Karan, L.S., Shopenskaya, T.A., Platonova, O.V., Zhuravlev, V.I., 2008. Epidemiology of West Nile infection in Volgograd, Russia, in relation to climate change and mosquito (Diptera: Culicidae) bionomics. Parasitol. Res. 103, 45–53. https://doi.org/10.1007/s00436-008-1050-0.

Platonov, A.E., Tolpin, V.A., Gridneva, K.A., Titkov, A.V., Platonova, O.V., Kolyasnikova, N.M., Busani, L., Rezza, G., 2014. The incidence of West Nile disease in Russia in relation to climatic and environmental factors. Int. J. Environ. Res. Pu 11, 1211–1232. https://doi.org/10.3390/ijerph110201211.

Pol, G.D., Crotta, M., Taylor, R.A., 2022. Modelling the temperature suitability for the risk of West Nile virus establishment in European Culex pipiens populations. Transbound. Emerg. Dis. https://doi.org/10.1111/tbed.14513.

Pruzanski, W., Altman, R., 1962. Encephalitis due to West Nile fever virus. World Neurol. 3, 524–528.

Reisen, W.K., Wheeler, S.S., 2019. Overwintering of West Nile virus in the United States. J. Med. Entomol. 56, 1498–1507. https://doi.org/10.1093/jme/tjz070.

Reisen, W.K., Fang, Y., Martinez, V.M., 2005. Avian host and mosquito (Diptera: Culicidae) vector competence determine the efficiency of West Nile and St. Louis encephalitis virus transmission. J. Med. Entomol. 42, 367–375. https://doi.org/10.1093/jmedent/42.3.367.

Reisen, W.K., Fang, Y., Martinez, V.M., 2006. Effects of temperature on the transmission of West Nile virus by *Culex tarsalis* (Diptera: Culicidae). J. Med. Entomol. 43, 309–317. https://doi.org/10.1603/0022-2585(2006)043[0309:eotott]2.0.co;2.

Riccò, M., Peruzzi, S., Balzarini, F., 2021. Epidemiology of West Nile virus infections in humans, Italy, 2012–2020: a summary of available evidences. Trop. Med. Infect Dis 6, 61. https://doi.org/10.3390/tropicalmed6020061.

Richards, S.L., Lord, C.C., Pesko, K., Tabachnick, W.J., 2009. Environmental and biological factors influencing Culex pipiens quinquefasciatus say (Diptera: Culicidae) vector competence for Saint Louis encephalitis virus. Am. J. Trop. Med. Hyg. 81, 264–272. https://doi.org/10.4269/ajtmh.2009.81.264.

Robertson, S.L., Caillouët, K.A., 2016. A host stage-structured model of enzootic West Nile virus transmission to explore the effect of avian stage-dependent exposure to vectors. J. Theor. Biol. 399, 33–42. https://doi.org/10.1016/j.jtbi.2016.03.031.

Rochlin, I., Harding, K., Ginsberg, H.S., Campbell, S.R., 2008. Comparative analysis of distribution and abundance of West Nile and eastern equine encephalomyelitis virus vectors in Suffolk County, New York, using human population density and land use/cover data. J. Med. Entomol. 45, 563–571. https://doi.org/10.1093/jmedent/45.3.563.

Rocklöv, J., Dubrow, R., 2020. Climate change: an enduring challenge for vector-borne disease prevention and control. Nat. Immunol. 21, 479–483. https://doi.org/10.1038/s41590-020-0648-y.

Rodrigues, L.R., McDermott, H.A., Villanueva, I., Djukarić, J., Ruf, L.C., Amcoff, M., Snook, R.R., 2022. Fluctuating heat stress during development exposes reproductive costs and putative benefits. J. Anim. Ecol. 91, 391–403. https://doi.org/10.1111/1365-2656.13636.

Root, J.J., Oesterle, P.T., Nemeth, N.M., Klenk, K., Gould, D.H., McLean, R.G., Clark, L., Hall, J.S., 2006. Experimental infection of fox squirrels (Sciurus niger) with West Nile virus. Am. J. Trop. Med. Hyg. 75, 697–701.

Rosà, R., Marini, G., Bolzoni, L., Neteler, M., Metz, M., Delucchi, L., Chadwick, E.A., Balbo, L., Mosca, A., Giacobini, M., Bertolotti, L., Rizzoli, A., 2014. Early warning of West Nile virus mosquito vector: climate and land use models successfully explain phenology and abundance of Culex pipiens mosquitoes in North-Western Italy. Parasit Vectors 7, 269. https://doi.org/10.1186/1756-3305-7-269.

Rudolf, I., Betášová, L., Blažejová, H., Venclíková, K., Straková, P., Šebesta, O., Mendel, J., Bakonyi, T., Schaffner, F., Nowotny, N., Hubálek, Z., 2017. West Nile virus in overwintering mosquitoes, Central Europe. Parasit Vectors 10, 452. https://doi.org/10.1186/s13071-017-2399-7.

Rueda, L.M., Patel, K.J., Axtell, R.C., Stinner, R.E., 1990. Temperature-dependent development and survival rates of Culex quinquefasciatus and Aedes aegypti (Diptera: Culicidae). J. Med. Entomol. 27, 892–898. https://doi.org/10.1093/jmedent/27.5.892.

Ruiz, M.O., Chaves, L.F., Hamer, G.L., Sun, T., Brown, W.M., Walker, E.D., Haramis, L., Goldberg, T.L., Kitron, U.D., 2010. Local impact of temperature and precipitation on West Nile virus infection in Culex species mosquitoes in Northeast Illinois, USA. Parasit Vectors 3, 19. https://doi.org/10.1186/1756-3305-3-19.

Ruybal, J.E., Kramer, L.D., Kilpatrick, A.M., 2016. Geographic variation in the response of Culex pipiens life history traits to temperature. Parasit Vectors 9, 116. https://doi.org/10.1186/s13071-016-1402-z.

Sadeghieh, T., Waddell, L.A., Ng, V., Hall, A., Sargeant, J., 2020. A scoping review of importation and predictive models related to vector-borne diseases, pathogens, reservoirs, or vectors (1999–2016). Plos One 15, e0227678. https://doi.org/10.1371/journal.pone.0227678.

Saiz, J.-C., 2020. Animal and human vaccines against West Nile virus. Pathogens 9, 1073. https://doi.org/10.3390/pathogens9121073.

Samy, A.M., Elaagip, A.H., Kenawy, M.A., Ayres, C.F.J., Peterson, A.T., Soliman, D.E., 2016. Climate change influences on the global potential distribution of the mosquito Culex quinquefasciatus, vector of West Nile virus and lymphatic filariasis. Plos One 11, e0163863. https://doi.org/10.1371/journal.pone.0163863.

Schou, M.F., Kristensen, T.N., Kellermann, V., Schlötterer, C., Loeschcke, V., 2014. A Drosophila laboratory evolution experiment points to low evolutionary potential under increased temperatures likely to be experienced in the future. J. Evol. Biol. 27, 1859–1868. https://doi.org/10.1111/jeb.12436.

Schrieke, H., Maignien, L., Constancias, F., Trigodet, F., Chakloute, S., Rakotoarivony, I., Marie, A., L'Ambert, G., Makoundou, P., Pages, N., Eren, A.M., Weill, M., Sicard, M., Reveillaud, J., 2021. The mosquito microbiome includes habitat-specific but rare symbionts. Comput. Struct. Biotechnol. J. 20, 410–420. https://doi.org/10.1016/j.csbj.2021.12.019.

Seah, A., Aik, J., Ng, L.-C., 2021. Effect of meteorological factors on Culex mosquitoes in Singapore: a time series analysis. Int. J. Biometeorol. 65, 963–965. https://doi.org/10.1007/s00484-020-02059-9.

Shaman, J., Day, J.F., Stieglitz, M., 2005. Drought-induced amplification and epidemic transmission of West Nile virus in southern Florida. J. Med. Entomol. 42, 134–141. https://doi.org/10.1093/jmedent/42.2.134.

Shaman, J., Harding, K., Campbell, S.R., 2011. Meteorological and hydrological influences on the spatial and temporal prevalence of West Nile virus in Culex mosquitoes, Suffolk County, New York. J. Med. Entomol. 48, 867–875. https://doi.org/10.1603/me10269.

Shartova, N., Mironova, V., Zelikhina, S., Korennoy, F., Grishchenko, M., 2022. Spatial patterns of West Nile virus distribution in the Volgograd region of Russia, a territory with long-existing foci. Plos Neglect. Trop. D 16, e0010145. https://doi.org/10.1371/journal.pntd.0010145.

Shelton, R.M., 1973. The effect of temperatures on development of eight mosquito species. Mosq. News 33, 1–12.

Shocket, M.S., Verwillow, A.B., Numazu, M.G., Slamani, H., Cohen, J.M., Moustaid, F.E., Rohr, J., Johnson, L.R., Mordecai, E.A., 2020. Transmission of West Nile and five other temperate mosquito-borne viruses peaks at temperatures between 23°C and 26°C. Elife 9. https://doi.org/10.7554/eLife.58511.

Smartt, C.T., Richards, S.L., Anderson, S.L., Vitek, C.J., 2010. Effects of forced egg retention on the temporal progression of West Nile virus infection in Culex pipiens quinquefasciatus (Diptera: Culicidae). Environ. Entomol. 39, 190–194. https://doi.org/10.1603/en09172.

Smith, K.H., Tyre, A.J., Hamik, J., Hayes, M.J., Zhou, Y., Dai, L., 2020. Using climate to explain and predict West Nile virus risk in Nebraska. GeoHealth 4, e2020GH000244. https://doi.org/10.1029/2020gh000244.

Smithburn, K., Hughes, T., Burke, A., Paul, J., 1940. A neurotropic virus isolated from the blood of a native of Uganda. Am. J. Trop. Med. 20, 471–472.

Soh, S., Aik, J., 2021. The abundance of Culex mosquito vectors for West Nile virus and other flaviviruses: a time-series analysis of rainfall and temperature dependence in Singapore. Sci. Total Environ. 754, 142420. https://doi.org/10.1016/j.scitotenv.2020.142420.

Soverow, J.E., Wellenius, G.A., Fisman, D.N., Mittleman, M.A., 2009. Infectious disease in a warming world: how weather influenced West Nile virus in the United States (2001–2005). Environ. Health Perspect. 117, 1049–1052. https://doi.org/10.1289/ehp.0800487.

Spanoudis, C.G., Andreadis, S.S., Tsaknis, N.K., Petrou, A.P., Gkeka, C.D., SavopoulouSoultani, M., 2018. Effect of temperature on biological parameters of the West Nile virus vector Culex pipiens form molestus (Diptera: Culicidae) in Greece: constant vs fluctuating temperatures. J. Med. Entomol. 56, 641–650. https://doi.org/10.1093/jme/tjy224.

Spigland, I., Jasinska-Klingberg, W., Hofshi, E., Goldbltjm, N., 1958. Clinical and laboratory observations in an outbreak of West Nile fever in Israel in 1957. Harefuah 54, 275–281.

Stilianakis, N.I., Syrris, V., Petroliagkis, T., Pärt, P., Gewehr, S., Kalaitzopoulou, S., Mourelatos, S., Baka, A., Pervanidou, D., Vontas, J., Hadjichristodoulou, C., 2016. Identification of climatic factors affecting the epidemiology of human West Nile virus infections in northern Greece. Plos One 11, e0161510. https://doi.org/10.1371/journal.pone.0161510.

Styer, L.M., Meola, M.A., Kramer, L.D., 2007. West Nile virus infection decreases fecundity of Culex tarsalis females. J. Med. Entomol. 44, 1074–1085. https://doi.org/10.1603/0022-2585%282007%2944%5b1074%3awnvidf%5d2.0.co%3b2.

Taylor, R., Hurlbut, H., 1953. Isolation of West Nile virus from Culex mosquitoes. J Egypt Med Assoc. 36, 199–208.

Taylor, R.M., Rizk, F., Work, T.H., Hurlbut, H.S., 1956. A study of the ecology of West Nile virus in Egypt 1. Am. J. Trop. Med. Hyg. 5, 579–620. https://doi.org/10.4269/ajtmh.1956.5.579.

Tempelis, C.H., 1975. REVIEW ARTICLE1: host-feeding patterns of mosquitoes, with a review of advances in analysis of blood meals by Serology2. J. Med. Entomol. 11, 635–653. https://doi.org/10.1093/jmedent/11.6.635.

Tran, A., Sudre, B., Paz, S., Rossi, M., Desbrosse, A., Chevalier, V., Semenza, J.C., 2014. Environmental predictors of West Nile fever risk in Europe. Int. J. Health Geogr. 13, 26. https://doi.org/10.1186/1476-072x-13-26.

Turell, M.J., Cohnstaedt, L.W., Wilson, W.C., 2020. Effect of environmental temperature on the ability of Culex tarsalis and Aedes taeniorhynchus (Diptera: Culicidae) to transmit Rift Valley fever virus. Vector-borne Zoonot 20, 454–460. https://doi.org/10.1089/vbz.2019.2554.

Ukawuba, I., Shaman, J., 2018. Association of spring-summer hydrology and meteorology with human West Nile virus infection in West Texas, USA, 2002–2016. Parasit Vectors 11, 224. https://doi.org/10.1186/s13071-018-2781-0.

Ulrich, J.N., Beier, J.C., Devine, G.J., Hugo, L.E., 2016. Heat sensitivity of wMel Wolbachia during Aedes aegypti development. Plos Neglect. Trop. D 10, e0004873. https://doi.org/10.1371/journal.pntd.0004873.

Valiakos, G., Papaspyropoulos, K., Giannakopoulos, A., Birtsas, P., Tsiodras, S., Hutchings, M.R., Spyrou, V., Pervanidou, D., Athanasiou, L.V., Papadopoulos, N., Tsokana, C., Baka, A., Manolakou, K., Chatzopoulos, D., Artois, M., Yon, L., Hannant, D., Petrovska, L., Hadjichristodoulou, C., Billinis, C., 2014. Use of wild bird surveillance, human case data and GIS spatial analysis for predicting spatial distributions of West Nile virus in Greece. Plos One 9, e96935. https://doi.org/10.1371/journal.pone.0096935.

Valzania, L., Martinson, V.G., Harrison, R.E., Boyd, B.M., Coon, K.L., Brown, M.R., Strand, M.R., 2018. Both living bacteria and eukaryotes in the mosquito gut promote growth of larvae. Plos Neglect. Trop. D 12, e0006638. https://doi.org/10.1371/journal.pntd.0006638.

Villena, O.C., Momen, B., Sullivan, J., Leisnham, P.T., 2018. Effects of ultraviolet radiation on metabolic rate and fitness of Aedes albopictus and Culex pipiens mosquitoes. Peerj 6, e6133. https://doi.org/10.7717/peerj.6133.

Vinogradova, E.B., 2000. *Culex pipiens* Pipiens Mosquitoes: Taxonomy, Distribution, Ecology, Physiology, Genetics, Applied Importance and Control. Pensoft Publishers.

Vogels, C.B.F., Fros, J.J., Göertz, G.P., Pijlman, G.P., Koenraadt, C.J.M., 2016. Vector competence of northern European Culex pipiens biotypes and hybrids for West Nile virus is differentially affected by temperature. Parasit Vectors 9, 393. https://doi.org/10.1186/s13071-016-1677-0.

Vogels, C.B., Göertz, G.P., Pijlman, G.P., Koenraadt, C.J., 2017. Vector competence of European mosquitoes for West Nile virus. Emerg Microbes Infec 6, e96. https://doi.org/10.1038/emi.2017.82.

Voskamp, A., Butchart, S.H.M., Baker, D.J., Wilsey, C.B., Willis, S.G., 2021. Site-based conservation of terrestrial bird species in the Caribbean and central and South America under climate change. Front. Ecol. Evol. 9, 625432. https://doi.org/10.3389/fevo.2021.625432.

Wang, G., Minnis, R.B., Belant, J.L., Wax, C.L., 2010. Dry weather induces outbreaks of human West Nile virus infections. BMC Infect. Dis. 10, 38. https://doi.org/10.1186/1471-2334-10-38.

Wang, J., Ogden, N.H., Zhu, H., 2011. The impact of weather conditions on Culex pipiens and Culex restuans (Diptera: Culicidae) abundance: a case study in Peel region. J. Med. Entomol. 48, 468–475. https://doi.org/10.1603/me10117.

Ward, M.P., Beveroth, T.A., Lampman, R., Raim, A., Enstrom, D., Novak, R., 2010. Field-based estimates of avian mortality from West Nile virus infection. Vector-borne Zoonot 10, 909–913. https://doi.org/10.1089/vbz.2008.0198.

Watanabe, K., Fukui, S., Ohta, S., 2017. Population of the temperate mosquito, Culex pipiens, decreases in response to habitat climatological changes in future. GeoHealth 1, 196–210. https://doi.org/10.1002/2017gh000054.

Watts, M., Kotsila, P., Mortyn, P.G., Monteys, V.S., 2021. The Rise of West Nile Virus in Europe: Investigating the Combined Effects of Climate, Land Use and Economic Changes. https://doi.org/10.21203/rs.3.rs-266617/v1.

Weingartl, H.M., Neufeld, J.L., Copps, J., Arszal, P.M., 2004. Experimental West Nile virus infection in blue jays (Cyanocitta cristata) and crows (Corvus brachyrhynchos). Vet. Pathol. 41, 362–370. https://doi.org/10.1354/vp.41-4-362.

Westbrook, C.J., Reiskind, M.H., Pesko, K.N., Greene, K.E., Lounibos, L.P., 2010. Larval environmental temperature and the susceptibility of Aedes albopictus Skuse (Diptera: Culicidae) to chikungunya virus. Vector-borne Zoonot 10, 241–247. https://doi.org/10.1089/vbz.2009.0035.

Wu, P., Yu, X., Wang, P., Cheng, G., 2019. Arbovirus lifecycle in mosquito: acquisition, propagation and transmission. Expert Rev. Mol. Med. 21, e1. https://doi.org/10.1017/erm.2018.6.

Zink, S.D., Slyke, G.A.V., Palumbo, M.J., Kramer, L.D., Ciota, A.T., 2015. Exposure to West Nile virus increases bacterial diversity and immune gene expression in Culex pipiens. Viruses 7, 5619–5631. https://doi.org/10.3390/v7102886.

Zouache, K., Fontaine, A., Vega-Rua, A., Mousson, L., Thiberge, J.-M., Lourenco-De-Oliveira, R., Caro, V., Lambrechts, L., Failloux, A.-B., 2014. Three-way interactions between mosquito population, viral strain and temperature underlying chikungunya virus transmission potential. Proc Royal Soc B Biol. Sci 281, 20141078. https://doi.org/10.1098/rspb.2014.1078.

Further reading

Chaulk, A.C., Carson, K.P., Whitney, H.G., Fonseca, D.M., Chapman, T.W., 2016. The arrival of the northern house mosquito Culex pipiens (Diptera: Culicidae) on Newfoundland's Avalon peninsula. J. Med. Entomol. 53, 1364–1369. https://doi.org/10.1093/jme/tjw105.

Chuang, T.-W., Knepper, R.G., Stanuszek, W.W., Walker, E.D., Wilson, M.L., 2011. Temporal and spatial patterns of West Nile virus transmission in Saginaw County, Michigan, 20032006. J. Med. Entomol. 48, 1047–1056. https://doi.org/10.1603/me10138.

Ewing, D.A., Purse, B.V., Cobbold, C.A., White, S.M., 2021. A novel approach for predicting risk of vector-borne disease establishment in marginal temperate environments under climate change: West Nile virus in the UK. J Roy Soc Interface 18, 20210049. https://doi.org/10.1098/rsif.2021.0049.

Pachka, H., Annelise, T., Alan, K., Power, T., Patrick, K., Véronique, C., Janusz, P., Ferran, J., 2016. Rift Valley fever vector diversity and impact of meteorological and environmental factors on Culex pipiens dynamics in the Okavango Delta, Botswana. Parasite Vector 9, 434. https://doi.org/10.1186/s13071-016-1712-1.

Paz, S., Albersheim, I., 2008. Influence of warming tendency on Culex pipiens population abundance and on the probability of West Nile fever outbreaks (Israeli case study: 2001–2005). Ecohealth 5, 40–48. https://doi.org/10.1007/s10393-007-0150-0.

Yoo, E.-H., Chen, D., Diao, C., Russell, C., 2016. The effects of weather and environmental factors on West Nile virus mosquito abundance in greater Toronto area. Earth Interact. 20, 1–22. https://doi.org/10.1175/ei-d-15-0003.1.